CW01186678

Extremely rare birds in the Western Palearctic

Marcel Haas

Extremely rare birds in the Western Palearctic

Marcel Haas

Lynx

Cover: Yellow-bellied Sapsucker *Sphyrapicus varius*, 08 October 2007, Selfoss, Árnessýsla, Iceland (© Örn Óskarsson)

First Edition: February 2012

© **Lynx Edicions** – Montseny, 8, 08193 Bellaterra, Barcelona (Spain), www.lynxeds.com
© text: Marcel Haas
© photographs: credited photographers

Design and layout: Elena Fonts Circuns

Printed by: Ingoprint, S.A.
Legal Deposit: B-1.578-2012
ISBN: 978-84-96553-83-5

All rights reserved. No form of reproduction, distribution, public communication or transformation of this work may be carried out without the authorization of its copyrights holders, except that foreseen by the law. Those needing to photocopy or electronically scan any part of this work should contact Lynx Edicions.

CONTENTS

Introduction 11
   Goal 11
   Region 11
   Period 11
   Records 12
   Species accounts 12
   Taxonomy and English names 12
   Acknowledgements 12

Species accounts 17
   DUCKS Anatidae 18
      Lesser Whistling Duck *Dendrocygna javanica* 18
      Ross's Goose *Anser rossii* 19
      Spur-winged Goose *Plectropterus gambensis* 23
      Cotton Pygmy Goose *Nettapus coromandelianus* 24
      Southern Pochard *Netta erythrophthalma* 24
      Redhead *Aythya americana* 25
      Spectacled Eider *Somateria fischeri* 28
      White-winged Scoter *Melanitta deglandi* 29
      Cape Teal *Anas capensis* 32
      Red-billed Teal *Anas erythrorhyncha* 32

LOONS Gaviidae 33
   Pacific Loon *Gavia pacifica* 33
ALBATROSSES Diomedeidae 35
   Atlantic Yellow-nosed Albatross *Thalassarche chlororhynchos* 35
   Shy Albatross *Thalassarche cauta* 37
   Tristan Albatross *Diomedea dabbenena* 38
PETRELS Procellariidae 39
   Southern Giant Petrel *Macronectes giganteus* 39
   Cape Petrel *Daption capense* 40
   Trindade Petrel *Pterodroma arminjoniana* 41
   Soft-plumaged Petrel *Pterodroma mollis* 42
   Bermuda Petrel *Pterodroma cahow* 42
   Black-capped Petrel *Pterodroma hasitata* 43
   Atlantic Petrel *Pterodroma incerta* 44
   Streaked Shearwater *Calonectris leucomelas* 45
   Flesh-footed Shearwater *Puffinus carneipes* 46
   Wedge-tailed Shearwater *Puffinus pacificus* 46
   Tropical Shearwater *Puffinus bailloni* 47
TROPICBIRDS Phaethontidae 48
   White-tailed Tropicbird *Phaethon lepturus* 48
GANNETS Sulidae 49
   Red-footed Booby *Sula sula* 49
   Masked Booby *Sula dactylatra* 50
FRIGATEBIRDS Fregatidae 52
   Ascension Frigatebird *Fregata aquila* 52
   Lesser Frigatebird *Fregata ariel* 53
HERONS Ardeidae 54
   Least Bittern *Ixobrychus exilis* 54
   Von Schrenck's Bittern *Ixobrychus eurhythmus* 56
   Dwarf Bittern *Ixobrychus sturmii* 57
   Indian Pond Heron *Ardeola grayii* 58
   Little Blue Heron *Egretta caerulea* 59
   Tricolored Heron *Egretta tricolor* 60
   Black Heron *Egretta ardesiaca* 61
   Black-headed Heron *Ardea melanocephala* 61
STORKS Ciconiidae 62
   Marabou Stork *Leptoptilos crumeniferus* 62
HAWKS Accipitridae 63
   Swallow-tailed Kite *Elanoides forficatus* 63
   African Fish Eagle *Haliaeetus vocifer* 64
   Bald Eagle *Haliaeetus leucocephalus* 64

Hooded Vulture *Necrosyrtes monachus*	65
FALCONS Falconidae	66
American Kestrel *Falco sparverius*	66
Amur Falcon *Falco amurensis*	68
RAILS Rallidae	70
Striped Crake *Aenigmatolimnas marginalis*	70
African Crake *Crex egregia*	72
Lesser Moorhen *Gallinula angulata*	74
CRANES Gruidae	75
Sandhill Crane *Grus canadensis*	75
PLOVERS Charadriidae	78
Three-banded Plover *Charadrius tricollaris*	78
Oriental Plover *Charadrius veredus*	79
Black-headed Lapwing *Vanellus tectus*	80
SANDPIPERS Scolopacidae	81
Long-toed Stint *Calidris subminuta*	81
Swinhoe's Snipe *Gallinago megala*	83
American Woodcock *Scolopax minor*	84
Hudsonian Godwit *Limosa haemastica*	85
Little Curlew *Numenius minutus*	88
Eskimo Curlew *Numenius borealis*	90
Grey-tailed Tattler *Tringa brevipes*	91
Willet *Tringa semipalmata*	92
SKUAS Stercorariidae	94
South Polar Skua *Stercorarius maccormicki*	94
GULLS Laridae	96
Brown-headed Gull *Croicocephalus brunnicephalus*	96
Relict Gull *Larus relictus*	96
Glaucous-winged Gull *Larus glaucescens*	97
Cape Gull *Larus dominicanus*	99
Slaty-backed Gull *Larus schistisagus*	101
TERNS Sternidae	102
Aleutian Tern *Onychoprion aleuticus*	102
Least Tern *Sternula antillarum*	103
Brown Noddy *Anous stolidus*	104
AUKS Alcidae	105
Long-billed Murrelet *Brachyrhamphus perdix*	105
Ancient Murrelet *Synthliboramphus antiquus*	107
Crested Auklet *Aethia cristatella*	108
Parakeet Auklet *Aethia psittacula*	109
Tufted Puffin *Fratercula cirrhata*	109

DOVES Columbidae..110
    Yellow-eyed Pigeon *Columba eversmanni*..............................110
    Mourning Dove *Zenaida macroura*..111
CUCKOOS Cuculidae..114
    Jacobin Cuckoo *Clamator jacobinus*..114
    Dideric Cuckoo *Chrysococcyx caprius*.....................................114
SWIFTS Apodidae...115
    Fork-tailed Swift *Apus pacificus*..115
    African Palm Swift *Cypsiurus parvus*.....................................117
ROLLERS Coraciidae...118
    Broad-billed Roller *Eurystomus glaucurus*..............................118
WOODPECKERS Picidae..119
    Northern Flicker *Colaptes auratus*...119
    Yellow-bellied Sapsucker *Sphyrapicus varius*.........................120
TYRANT-FLYCATCHERS Tyrannidae...122
    Eastern Phoebe *Sayornis phoebe*...122
    Acadian Flycatcher *Empidonax virescens*..............................123
    Alder Flycatcher *Empidonax alnorum*....................................124
    Least Flycatcher *Empidonax minimus*...................................125
    Fork-tailed Flycatcher *Tyrannus savana*................................125
VIREOS Vireonidae..126
    White-eyed Vireo *Vireo griseus*...126
    Yellow-throated Vireo *Vireo flavifrons*..................................127
    Philadelphia Vireo *Vireo philadelphicus*................................128
SHRIKES Laniidae...129
    Long-tailed Shrike *Lanius schach*...129
CROWS Corvidae..132
    Daurian Jackdaw *Corvus dauuricus*.......................................132
    Pied Crow *Corvus albus*..134
GOLDCRESTS Regulidae..135
    Ruby-crowned Kinglet *Regulus calendula*............................135
LARKS Alaudidae...136
    Chestnut-headed Sparrow-Lark *Eremopterix signatus*........136
    Hume's Short-toed Lark *Calandrella acutirostris*.................136
SWALLOWS Hirundinidae..137
    Banded Martin *Riparia cincta*..137
    Tree Swallow *Tachycineta bicolor*..138
    Purple Martin *Progne subis*..140
    Ethiopian Swallow *Hirundo aethiopica*................................141
LEAF WARBLERS Phylloscopidae...142
    Eastern Crowned Warbler *Phylloscopus coronatus*............142

CONTENTS

    Plain Leaf Warbler *Phylloscopus neglectus* 144
GRASSHOPPER WARBLERS Locustellidae 145
    Gray's Grasshopper Warbler *Locustella fasciolata* 145
REED WARBLERS Acrocephalidae 146
    Thick-billed Warbler *Iduna aedon* 146
    Oriental Reed Warbler *Acrocephalus orientalis* 148
WAXWINGS Bombycillidae 149
    Cedar Waxwing *Bombycilla cedrorum* 149
NUTHATCHES Sittidae 151
    Red-breasted Nuthatch *Sitta canadensis* 151
MOCKINGBIRDS Mimidae 154
    Northern Mockingbird *Mimus polyglottos* 152
    Brown Thrasher *Toxostoma rufum* 154
    Grey Catbird *Dumetella carolinensis* 155
STARLINGS Sturnidae 157
    Daurian Starling *Agropsar sturninus* 157
SUNBIRDS Nectariniidae 159
    Purple Sunbird *Cinnyris asiaticus* 159
FLYCATCHERS Muscicapidae 160
    Varied Thrush *Ixoreus naevius* 160
    Wood Thrush *Hylocichla mustelina* 161
    Veery *Catharus fuscescens* 162
    Tickell's Thrush *Turdus unicolor* 164
    Asian Brown Flycatcher *Muscicapa dauurica* 165
    Rufous-tailed Robin *Luscinia sibilans* 167
    Siberian Blue Robin *Luscinia cyane* 168
    Daurian Redstart *Phoenicurus auroreus* 170
    Ant Chat *Myrmecocichla aethiops* 171
    Variable Wheatear *Oenanthe picata* 171
    Mugimaki Flycatcher *Ficedula mugimaki* 171
WAGTAILS Motacillidae 172
    Forest Wagtail *Dendronanthus indicus* 172
    Amur Wagtail *Motacilla leucopsis* 173
FINCHES Fringillidae 174
    Evening Grosbeak *Hesperiphona vespertina* 174
BUNTINGS Emberizidae 176
    Summer Tanager *Piranga rubra* 176
    Dickcissel *Spiza americana* 177
    Eastern Towhee *Pipilo erythrophthalmus* 177
    Lark Sparrow *Chondestes grammacus* 178
    Savannah Sparrow *Passerculus sandwichensis* 179

Fox Sparrow *Passerella iliaca* 181
Chestnut-eared Bunting *Emberiza fucata* 182
Yellow-browed Bunting *Emberiza chrysophrys* 183
Chestnut Bunting *Emberiza rutila* 185
Brown-headed Cowbird *Molothrus ater* 188
Yellow-headed Blackbird *Xanthocephalus xanthocephalus* 189
Golden-winged Warbler *Vermivora chrysoptera* 190
Blue-winged Warbler *Vermivora pinus* 191
Tennessee Warbler *Vermivora peregrina* 192
Chestnut-sided Warbler *Dendroica pensylvanica* 193
Cerulean Warbler *Dendroica cerulea* 194
Black-throated Blue Warbler *Dendroica caerulescens* 195
Black-throated Green Warbler *Dendroica virens* 197
Blackburnian Warbler *Dendroica fusca* 198
Cape May Warbler *Dendroica tigrina* 199
Magnolia Warbler *Dendroica magnolia* 200
Palm Warbler *Dendroica palmarum* 202
Bay-breasted Warbler *Dendroica castanea* 203
Ovenbird *Seiurus aurocapilla* 203
Louisiana Waterthrush *Seiurus motacilla* 206
Hooded Warbler *Wilsonia citrina* 207
Wilson's Warbler *Wilsonia pusilla* 208
Canada Warbler *Wilsonia canadensis* 208

References 211

Appendix 237
New species since 2009 237

Index 239

INTRODUCTION

The present work is the result of my lifelong interest in the occurrence of rare birds in the Western Palearctic (WP). In 2002, my brother Lucien Davids and I started the now defunct website *wpbirds.com*. Indeed, listing all records of all rare birds in the WP proved to be very difficult and time-consuming and in 2009 we decided to give up the website project. Meanwhile, I had started to collect records of the rarest WP birds. Gradually, I developed the idea of compiling a book that concentrates on birds only very rarely occurring in the WP.

GOAL

The book's main goal is to summarize all available data on records of extremely rare birds in the WP. Consequently, detailed listings of records and references form the basis of the book. To document the records, I have tried to include one photograph of the bird (or specimen) in question.

REGION

All included records are from within the boundaries as defined by *The Birds of the Western Palearctic* (Cramp *et al.* 1977–94; Snow & Perrins 1998). These boundaries (especially the southern one) are rather arbitrary and several authors have expressed the view that the Arabian Peninsula should be included in the WP (i.e. Martins & Hirschfeld 1998; Roselaar 2006). For mainly practical reasons I have decided not to adopt this view.

PERIOD

The covered period is 1800–2008. In the species accounts, records from 2009 to August 2010 are also included for the sake of completeness. This also applies to three new species to the WP (all three photographically documented and listed in the appendix).

RECORDS

All species recorded less than 10 times in the WP are listed. Only records accepted by the relevant rarities committees are included and have to have been accepted on the respective national lists (i.e. category A or B). Thus, records of introduced species (category C) or certain escapes (category E) are not included. For countries without a rarities or records committee, publications on the occurrence of rare vagrants are referenced.

In several countries, some records are placed in category D. This category is meant for a species for which it is unclear whether it is a genuine vagrant (categories A and B) or an obvious escape (category E). I have only listed category A and B records. Category D records are mentioned in the species accounts. Two species have been recorded in the WP only as category D species (Swainson's Hawk *Buteo swainsoni* in France and Norway, and Violet-backed Starling *Cinnyricinclus leucogaster* in Israel).

Details of specimens are included as well. Abbreviations of the museums where the specimens are kept are listed in Table 1. These are used in the record listings, together with the museum registration number, if available.

SPECIES ACCOUNTS

The species accounts start with a paragraph on distribution and movements. The second paragraph details the records that have occurred during the covered period.

The records of some species are analysed, but I have refrained from doing so when only very few records are available. As many Nearctic vagrants have been recorded in September–October, I have refrained from analysing most of them as well. For species recorded in Britain, the excellent book by Slack (2009) and the very interesting series of articles by Elkins (1979, 1988, 1999, 2008) should be consulted. Finally, Vinicombe & Cottridge (1996) and McLaren *et al.* (2006) are highly recommended as these include thought-provoking discussions on the occurrence of rare birds.

TAXONOMY AND ENGLISH NAMES

The taxonomic decisions by the *Commissie Systematiek Nederlandse Avifauna* (CSNA) are followed (cf. Sangster *et al.* 1999, 2003, 2009). This also applies to the editorial announcements published in each first issue of *Dutch Birding* since 2001. The English names given by Gill & Donsker (2010) are used.

ACKNOWLEDGEMENTS

First, I would like to thank Pierre-André Crochet – with whom I share the same avifaunistic interest – for co-authoring articles updating the WP list (published in *Dutch Birding*); and for our very stimulating discussions on this and related subjects. Second, I should thank both Arnoud van den Berg and

INTRODUCTION

Cecilia Bosman for their pleasant company during the many times I visited the *Dutch Birding* library. Arnoud and I also had many very stimulating discussions.

Thanks are also due to the following people for their help in checking records and collecting background information: Jochen Dierschke, Gonçalo Elias, Lee Evans, Giancarlo Fracasso, Geert Groot Koerkamp, Andrew Harrop, Kees Hazevoet, João Jara, Frédéric Jiguet, Yann Kolbeinsson, Paul Milne, Gunnlaugur Pétursson, Viggo Ree and Nir Sapir.

Furthermore, I should thank: Ian Andrews, Chris Batty, Max Berlijn, David Boertmann, Colin Bradshaw, Joël Bried, Graham Bundy, Vegard Bunes, Agris Celmiņš, Giuliano Doria, Philippe Dubois, Joris Elst, Bruria Gal, Ramond Galea, George Gregory, Tomas Haraldsson, Jan Lifjeld, Antero Lindholm, Dave McAdams, Peter Meininger, Nigel Monaghan, Killian Mullarney, Daniele Occhiato, Margus Ots, Charles Perez, Yoav Perlman, Nikos Probonas, Robert Prŷs-Jones, Dare Šere, Valentin Serebryakov, Hadoram Shirihai, Søren Sørensen, Torsten Stjernberg, Joe Sultana, Kasper Thorup and Arend Wassink.

Thanks also to the following people for giving permission to publish their often unique photographs: Peter Alfrey, Khaled Al-Ghanem, Pekka Alho, Keith Allsopp, Itzik Amir, Gert Anderson, Antonio Antonucci, Duarte Araújo, Steve Baines, Theo Bakker, Filipe Barata, Chris Batty, Arnoud van den Berg, Lars Bergersen, Peter Bergman, Daniel Bergmann, Jens Berkler, David Bigas, Peter Bjurenstål, Giovanni Boano, J Borg, Enrico Borgo, Morten Brendstrup-Hansen, Joël Bried, Tony Broome, Jean Brosset, Ruud Brouwer, Dave Bryan, Tony Byar, Ingvar Byrkjedal, Richard Campey, Christian Cederroth, Niels Hesselbjerg Christensen, Jim Clift, Paul Condon, Bill Condry, Guy Conrady, Andrea Corso, David Cottridge, Tony Croucher, Lucien Davids, Gary Davies, Adriano De Faveri, Kris De Rouck, Wendy Dickson, Jochen Dierschke, Marten van Dijl, Gonçalo Elias, Joakim Engel, Manuel Fernandes, Ian Fisher, Þórhallur Frímannsson, Göran Frisk, Raymond Galea, Steve Gantlett, Antonio Hernández García, David Godfrey, Y Golan, Stefan Golaszewsky, Lee Gregory, Atle Grimsby, Klaus Günther, Nic Hallam, Hugh Harrop, Margaret Hart (American Museum of Natural History), Paul Harvey, Vidar Heibo, Mikko Heikkinen, Jens Hering, Jóhann Óli Hilmarsson, J F Holloway, Oz Horine, Tero Ilomäki, Gary Jenkins, Frédéric Jiguet, John Johns, Andrew Jones (The Cleveland Museum of Natural History), Vytautas Jusys, Jan van de Kam, Paul and Andrea Kelly, John Kinsley, Peter Knaus, Yann Kolbeinsson, Hans Krabsen, Fredrik Kræmer, Ole Krogh, Tomasz Kulakowski, Jan van der Laan, Vincent Legrand, Bud Lehnhausen, Harry Lehto, Ernest Glyndwr Lewis, Jan Lifjeld, Lindblad Expeditions, Sture Lindholm, Y Livneh, Roger Long, Ferran López, Juan Antonio Lorenzo, Tim Loseby, Vladimir Loskot, Pete Loud, Fabio Lo Valvo, D Lukasik,

Arne Lundberg, Micky Maher, Mathias Mähler, Dan Mangsbo, Steve Mann, Amanda Martin, Bruno Massa, Anthony McGeehan, Ross McGregor, Clive McKay, Eric Meek, Richard Mills, Steve Minton, Dominic Mitchell, K Møklegård, Killian Mullarney, Nikolay Neyfeld, János Oláh, Silas Olofson, Björn Olsen, Knut Olsen, Lasse Olsen, Troels Eske Ortvad, Örn Óskarsson, Phil Palmer, John Parrot, Petri Pelttari, Carlos Pereira, Yoav Perlman, Nick Picozzi, Mike Pope, Stefan Pfützke, Lars Maltha Rasmussen, Viggo Ree, Ian Robertson, Steffan Rodebrand, Angela Ross, René van Rossum, Petri Salo, John Sanders, Ludovic Scalabre, Hanno Schäfer, Martin Scott, Martin Semisch, Dare Šere, Hadoram Shirihai, Ingvar A Sigurðsson, Søren Sørensen, Jostein Sørgård, Silke Sottorf, Laurent Spanneut, Kristian Stahl, Lisa Steiner, Harald Støen, Jan van der Straaten, Pedro López Suárez, Joe Sultana, Julian Sykes, Harri Taavetti, A R Taylor, Adi Teitler, David Tipling, Böðvar Þórisson, Skarphéðinn Þórisson, Gunnlaugur Þráinsson, Ingvar Torsson, Nigel Tucker, Morten Vang, Sander van de Water, Rob Wilson and Steve Young.

A number of people helped me enormously in collecting unpublished or hard-to-get photographs. In particular, I thank Ian Andrews for sending the photographs published in Forrester *et al.* (2007); Viggo Ree for tracking down several unpublished Norwegian photographs; Hans Bister for sending scans of Swedish rarities published in *Roadrunner* or on the *Club300* website; and Steve Gantlett and André van Loon for sending photographs published in *Birding World* and *Dutch Birding*, respectively.

Gerald Oreel kindly edited the text and did a wonderful job. Gert Ottens and Peter de Vries kindly read drafts of the manuscript and provided useful feedback.

Finally, I must thank my family for their continuous support and interest throughout the project.

Table 1 MUSEUM ABBREVIATIONS USED IN SPECIES ACCOUNTS.

AMNH, American Museum of Natural History, New York, New York, USA
BDGC, Biology Department, University of Las Palmas de Gran Canaria, Canary Islands, Spain
BGM, Beit Gordon Museum, Degania Alef, Israel
BM, Bergen Museum, Bergen, Norway
CM, Castle Museum, Norwich, Norfolk, England, UK
CMNH, Cleveland Museum of Natural History, Cleveland, Ohio, USA
FMNH, Finnish Museum of Natural History, Helsinki, Finland
FMC, The Field Museum, Chicago, Illinois, USA
ICPL, Instituto Cabrera Pinto de La Laguna, Tenerife, Canary Islands, Spain
IMNH, Icelandic Museum of Natural History, Reykjavík, Iceland
INFS, Istituto Nazionale per la Fauna Selvatica, Ozzano dell'Emilia, Bologna, Italy
ISM, Isles of Scilly Museum, St Mary's, Isles of Scilly, England, UK
IVVH, Institut für Vogelforschung "Vogelwarte Helgoland", Wilhelmshaven, Niedersachsen, Germany
MCGD, Museo Civico di Storia Naturale "Giacomo Doria", Genova, Italy
MCM, Museu Carlos Machado, Ponta Delgada, São Miguel, Azores, Portugal
MCNT, Museo de Ciencias Naturales de Tenerife, Tenerife, Canary Islands, Spain
MDNL, Museum d'Histoire Naturelle, Lille, Nord, France
MJBF, Museu do Jardim Botânico do Funchal, Funchal, Madeira, Portugal
MM, Manx Museum, Isle of Man, UK
MNHN, Muséum National d'Histoire Naturelle, Paris, France
MNHT, Museo de la Naturaleza y el Hombre de Tenerife, Tenerife, Canary Islands, Spain
MRSN, Museo Regionale di Scienze Naturali, Torino, Italy
MSNT, Museo di Storia Naturale, Terrasini, Sicily, Italy
MT, Muséum d'histoire naturelle, Toulouse, France
NHM, Natural History Museum, Tring, Hertfordshire, England, UK
NHMB, Natural History Museum, Basel, Switzerland
NMF, Naturwissenschaftliches Museum, Flensburg, Schleswig-Holstein, Germany
NMI, Natural Museum of Ireland – Natural History, Dublin, Ireland
NMNH, National Museum of Natural History, Mdina, Malta
NMS, National Museums of Scotland, Edinburgh, Scotland, UK
NNM, Nationaal Natuurhistorisch Museum, Leiden, Netherlands
SMNH, Swedish Museum of Natural History, Stockholm, Sweden
TAU, Tel Aviv University, Tel Aviv, Israel
UM, Ulster Museum, Belfast, Northern Ireland, UK
ZMO, Zoologisk Museum, Oslo, Norway
ZMSP, Zoological Museum, St Petersburg, Russia
ZMUC, Zoological Museum, University of Copenhagen, Denmark

SPECIES ACCOUNTS

DUCKS Anatidae (n=10)

Lesser Whistling Duck *Dendrocygna javanica*

Breeds in tropical Asia, from Pakistan and India in the west to China and much of south-eastern Asia in the east. Mostly resident but eastern Chinese birds move south in winter. Dispersive movements may occur in dry years (del Hoyo *et al.* 1992; Madge & Burn 1988). It has also occurred in Oman (Eriksen *et al.* 2003).

The single bird recorded in Israel was photographed and thought at the time to be a Plumed Whistling Duck *D. eytoni* and subsequently a Fulvous Whistling Duck *D. bicolor* until Hadoram Shirihai saw the photograph in 1991 during his research on Shirihai (1996) and correctly identified this bird. This is believed by Shirihai (1996) to be a genuine vagrant. One can indeed argue that it was a wandering bird because its range is not very far from Israel; as well the fact that it is in the right plumage for the time of year and that no Lesser Whistling Ducks were listed as being in captivity in any zoo in Israel at that time support genuine vagrancy.

1 record

Israel (1)
1 • 15 November 1966 to mid-March 1967, Ma'agan Mikhael [Shirihai 1996]

November 1966, Ma'agan Mikhael, Israel. © Y. Livneh

Ross's Goose *Anser rossii*

Breeds in Arctic Canada, mainly in Northwest Territories, and has recently expanded west to western and southern coasts of Hudson Bay. Winters in California, Arkansas, New Mexico and central-northern Mexico, and along the coasts of Louisiana and Texas (van den Berg 2004a; del Hoyo *et al*. 1992).

Van den Berg (2004a) mentioned more records from Europe (e.g. Britain, Denmark, Faeroes, Germany, Norway and Sweden), but these are all considered to relate to birds from captivity. In 2006–07, a mixed pair of Ross's and Barnacle Geese *Branta leucopsis* was found on Kolguev in the Barents Sea, Russia, and produced three hybrid young in 2007. The outcome of the 2007 brood was never determined, but four eggs were found during a later visit (Kondratyev & Zöckler 2009). A flock of four birds occurred in the Netherlands in September–October 2009, and in the winter of 2009/10 up to three were again present (van den Berg & Haas 2010abc; Ovaa *et al*. 2010).

January 1995, Stellendam, Goedereede, Zuid-Holland, Netherlands. © Marten van Dijl

14 January 2001, Nieuwendijk, Korendijk, Zuid-Holland, Netherlands. © Marten van Dijl

26 February 2000, Anjumerkolken, Dongeradeel, Friesland, Netherlands. © Lucien Davids

Ross's Goose *Anser rossii*

4 records (7 individuals)

Netherlands (4/7)

1. • 30 November 1985, Santpoort-Noord/Velserbroek, Velsen, Noord-Holland, adult, white morph [van den Berg & Bosman 2001; van den Berg & Cottaar 1986; Blankert et al. 1987; de By & Winkelman 1987]
   • 01 December 1985, Assendelft, Zaanstad, Noord-Holland, adult, white morph
2. • 20–27 January 1988, Middelplaten, Veerse Meer, Goes, Zeeland, *two* (adults), white morph [van den Berg et al. 1989, 1991, 1992, 1993; Berlijn 2004; de By et al. 1992, 1993; de By & CDNA 1991; de By & de Knijff 1989; van der Vliet et al. 2002; Wiegant et al. 1994ab, 1995, 1996ab, 1997]
   • 23 May 1988, Biervliet, Terneuzen, Zeeland, adult, white morph
   • 15 January 1989, Stellendam, Goedereede, Zuid-Holland, *two* (adults), white morph
   • 25 February to 11 March 1989, Wûnseradiel, Friesland, *two* (adults), white morph
   • 18 November 1989 to 19 February 1990, Stad aan 't Haringvliet, Middelharnis, Zuid-Holland & Stellendam, Goedereede, Zuid-Holland, *maximum of two* (adults), white morph
   • 01 December 1990 to 23 March 1991, Stellendam, Goedereede, Zuid-Holland, adult, white morph
   • 25 January to 16 March 1991, Workumerwaard, Nijefurd, Friesland, adult, white morph
   • 29 October to 31 December 1991, Stad aan 't Haringvliet, Middelharnis, Zuid-Holland & Stellendam, Goedereede, Zuid-Holland, adult, white morph
   • 15 February to 16 March 1992, Workumerwaard, Nijefurd, Friesland, adult, white morph
   • 30 December 1992, Stellendam, Goedereede, Zuid-Holland, adult, white morph
   • 09 January to 06 March 1994, Stellendam, Goedereede, Zuid-Holland & Slikken van Flakkee, Goedereede, Zuid-Holland, adult, white morph
   • February 1994, Workumerwaard, Nijefurd, Friesland, adult, white morph
   • 20 February 1994, Maasvlakte, Rotterdam, Zuid-Holland, adult, white morph
   • 13 January to 10 March 1995, Scheelhoek, Goedereede, Zuid-Holland, adult, white morph
   • 12 November 1995 to 06 March 1996, Plaat van Scheelhoek, Goedereede, Zuid-Holland, adult, white morph
3. • 14–15 November 1998, Workum, Nijefurd, Friesland & Oudega, Wymbritseradiel, Friesland, adult, white morph [Berlijn 2004; Ovaa et al. 2008, 2009; van der Vliet et al. 2001, 2002, 2003, 2004, 2005, 2006, 2007]
   • 25 February to 06 April 2000, Anjumerkolken, Dongeradeel, Friesland & Bandpolder, Dongeradeel, Friesland & Lauwersmeer, De Marne, Groningen, adult, white morph

- 17–19 October 2000, Marnewaard, Kollumerland en Nieuwskruisland, Friesland & Lauwersmeer, De Marne, Groningen, adult, white morph
- 19 November to 10 December 2000, Gaast, Wûnseradiel, Friesland & Piaam, Wûnseradiel, Friesland, adult, white morph
- 23 January to 14 February 2001, Workumerwaard, Nijefurd, Friesland & Doniaburen, Wûnseradiel, Friesland & Gaast, Wûnseradiel, Friesland & Ferwert, Ferwerderadiel, Friesland, adult, white morph
- 15 February to 18 March 2001, Anjumerkolken, Dongeradeel, Friesland & Jaap Deensgat, De Marne, Groningen, adult, white morph
- 01 March to 14 April 2002, Anjumerkolken, Dongeradeel, Friesland & Jaap Deensgat, De Marne, Groningen, adult, white morph
- 14–26 September 2002, Bandpolder, Dongeradeel, Friesland, adult, white morph
- 28 September to 18 October 2002, Workumerwaard, Nijefurd, Friesland & Gaast, Wûnseradiel, Friesland, adult, white morph
- 16 February 2003, Mirns, Gaasterlân-Sleat, Friesland, adult, white morph
- 17–18 February 2004, Ferwerderadeelsbuitenpolder, Ferwerderadiel, Friesland, adult, white morph
- 05–23 March 2004, Oostvaardersplassen, Almere/Lelystad, Flevoland & Polder IJdoorn, Durgerdam, Amsterdam, Noord-Holland, adult, white morph
- 13–27 April 2004, Ferwerderadeelsbuitenpolder, Ferwerderadiel, Friesland, *maximum of two* (adults), white morph
- 02 May to 02 June 2004, Oostvaardersplassen, Almere, Flevoland, adult, white morph
- 07 May 2004, Ezumakeeg, Dongeradeel, Friesland, adult, white morph
- 19 December 2004, Hurwenensche Uiterwaarden, Rossum, Gelderland, adult, white morph
- 02–21 January 2005, Bochtjesplaat, Dongeradeel, Friesland & Anjumerkolken, Dongeradeel, Friesland, adult, white morph
- 02 February to 18 April 2005, Oostvaardersplassen, Almere/Lelystad, Flevoland & Polder Zeevang, Zeevang, Noord-Holland & Oosterpoel, Waterland, Noord-Holland, adult, white morph
- 27 April 2005, Jan Durkspolder, Oudega, Smallingerland, Friesland, adult, white morph
- 19–22 October 2005, Anjumerkolken, Dongeradeel, Friesland, adult, white morph
- 25 November to 29 December 2005, Mirns, Gaasterlân-Sleat, Friesland & Oudermirdum, Gaasterlân-Sleat, Friesland, adult, white morph
- 26 March, 02 April and 20 April 2006, Dollardkwelder, Reiderland, Groningen, adult, white morph
- 19–22 January 2007, Schildmeer, Slochteren, Groningen, adult, white morph
- 25 March to 05 April 2007, Ferwâlde (Ferwoude), Wûnseradiel,

Ross's Goose *Anser rossii*

Friesland, adult, white morph
4 • 20 November 1999 to 31 May 2000, Korendijkse Slikken, Korendijk, Zuid-Holland & Strijen, Zuid-Holland & Scheelhoek, Goedereede, Zuid-Holland & Stad aan 't Haringvliet, Middelharnis, Zuid-Holland & Slijkplaat, Middelharnis, Zuid-Holland, adult, white morph [Berlijn 2004; Ovaa et al. 2008; van der Vliet et al. 2001, 2002, 2003, 2004, 2005, 2006, 2007]
• 31 August 2000 to 27 April 2001, Korendijkse Slikken, Korendijk, Zuid-Holland & Nieuwendijk, Korendijk, Zuid-Holland & Stellendam, Goedereede, Zuid-Holland & Plaat van Scheelhoek, Goedereede, Zuid-Holland & Tonnekreek, Zevenbergen, Noord-Brabant & Willemstad, Zevenbergen, Noord-Brabant, *maximum of two* (adults), white morph
• 16 December 2001 to 07 April 2002, Den Bommel, Oostflakkee, Zuid-Holland & Plaat van Scheelhoek, Goedereede, Zuid-Holland & Stellendam, Goedereede, Zuid-Holland & Strijen, Zuid-Holland & Rammegors, Tholen, Zeeland, adult, white morph
• 23 November 2002 to 03 January 2003, Korendijkse Slikken, Korendijk, Zuid-Holland & Plaat van Scheelhoek, Goedereede, Zuid-Holland & Stinkgat, Tholen, Zeeland, adult, white morph
• 14 January to 27 April 2003, Plaat van Scheelhoek, Goedereede, Zuid-Holland, adult, white morph
• 17 January to 12 February 2003, Stinkgat, Tholen, Zeeland & Rammegors, Tholen, Zeeland, adult, white morph
• 19 May to 02 June 2003, Slijkplaat, Hellevoetsluis, Zuid-Holland & Haringvliet, Oostflakkee, Zuid-Holland, adult female, white morph, at nest (unpaired), five eggs
• 25 December 2003 to 08 April 2004, Korendijkse Slikken, Korendijk, Zuid-Holland & Slikken van Flakkee, Dirksland, Zuid-Holland & Ouddorp, Goedereede, Zuid-Holland & Rammegors, Tholen, Zeeland & Scherpenissepolder, Tholen, Zeeland & Stinkgat, Tholen, Zeeland & Yerseke, Reimerswaal, Zeeland, adult, white morph
• 16 October 2004 to 21 February 2005, Slikken van Flakkee, Dirksland, Zuid-Holland & Ouddorp, Goedereede, Zuid-Holland & Rammegors, Tholen, Zeeland & Scherpenissepolder, Tholen, Zeeland & Yerseke, Reimerswaal, Zeeland, adult, white morph
• 15 March to 29 April 2005, Scherpenissepolder, Tholen, Zeeland & Goedereede, Zuid-Holland & Westplaat Buitengronden, Melissant, Dirksland, Zuid-Holland, *maximum of two* (adults), white morph
• 28 January to 02 May 2006, Yerseke Moer, Reimerswaal, Zeeland & Scherpenissepolder, Tholen, Zeeland & Oudeland van Strijen, Strijen, Zuid-Holland, adult, white morph
• 28 January to 12 February 2007, Yerseke Moer, Kapelle, Zeeland, adult, white morph

Spur-winged Goose *Plectropterus gambensis*

Breeds in sub-Saharan Africa, from Gambia to Ethiopia and south to southern Africa (del Hoyo *et al.* 1992; Madge & Burn 1988).

The Egyptian reports in Cramp & Simmons (1977) probably concern escaped birds (cf. Crochet & Spaans 2008; Goodman & Meininger 1989).

2 records (15 individuals)

Mauritania (1/14)
1 • 11–14 December 2004, Ebel Kheaiznaya, Banc d'Arguin, *maximum of 14* [Crochet & Spaans 2008; Isenmann 2007]

Morocco (1)
1 • 12 and 23 March 1984, Souss, immature [Bouwman 1985; Thévenot *et al.* 2003]
• 23 April to 12 December 1984, Oued Massa, immature

12 March 1984, Souss, Morocco. © Morten Brendstrup-Hansen

December 2004, Ebel Kheaiznaya, Banc d'Arguin, Mauritania. © Jan van de Kam

Cotton Pygmy Goose *Nettapus coromandelianus*

Breeds in south-eastern Asia, from India east through Burma to China and south to Malaysia and Indonesia; also in Australia. Largely resident but Chinese birds move south in winter. Occurs as a vagrant in the Middle East where birds have turned up in Oman, UAE and Bahrain (Eriksen *et al.* 2003; del Hoyo *et al.* 1992; Madge & Burn 1988; Pedersen & Aspinall 2010).

The record from Jordan was the first proof of genuine occurrence in the Western Palearctic. Two females found at a market in Ashar, Iraq, in 1976 were said to have been trapped at Hawr al Hammar the previous day but the origin of these birds is uncertain (Kainady 1976).

1 record

Jordan (1)
1 • 09–10 April 1997, Aqaba sewage works, female [Andrews *et al.* 1999; Bashford 1997]

Southern Pochard *Netta erythrophthalma*

Breeds in Africa, from South Africa north-east to Ethiopia and in South America (where very rare, perhaps now only breeding in Brazil and Venezuela). Largely resident but some birds show northward movements in winter (ringing recoveries have shown birds to occur as far north as Kenya in winter) (del Hoyo *et al.* 1992; Madge & Burn 1988).

The bird in Israel was probably a genuine vagrant.

1 record

Israel (1)
1 • 22 April to 08 May 1998, Eilat [Sapir & Israeli Rarities and Distribution Committee 2007; Shirihai 2000; Smith *et al.* 2001]

April 1998, Eilat, Israel. © Hadoram Shirihai

Redhead *Aythya americana*

Breeds on prairie marshes in western Canada and USA, from central Canada south to California, New Mexico and Nebraska. Winters chiefly from British Columbia and Great Lakes region south to central Mexico and Cuba (del Hoyo *et al*. 1992; Madge & Burn 1988). It was also recorded in Greenland in June 1953 (Boertmann 1994).

A record in Germany in April–June 2000 was placed in category D (Deutsche Seltenheitenkommission 2006). All British records are now under review, including a record in January 2006 on Lewis, Outer Hebrides, which has not been formally accepted (yet) (Chris Batty in litt.).

6 records

Britain (3)
1. • 08–27 March 1996, Bleasby, Nottinghamshire, male [British Ornithologists' Union 1998; Dennis 1996, 1998; Rogers & Rarities Committee 1997, 1998]
   • 04–24 February 1997, Rutland Water, Leicestershire, male
2. • 07 November 2001 to 05 February 2002, Kenfig Pool, Glamorgan, male [Bolt 2001; Rogers & Rarities Committee 2003, 2004, 2005]
   • 21 September 2002 to 31 January 2003, Kenfig Pool, Glamorgan, male
   • 06–16 February 2003, Cosmeston Lakes, Glamorgan, male
   • 16–23 February 2003, Cisvane Reservoir, Glamorgan, male

March 1996, Bleasby, Nottinghamshire, Britain.
© Steve Young

Extremely rare birds in the Western Palearctic

Redhead *Aythya americana*

February 2004, Barra, Outer Hebrides, Britain. © Martin Scott

10 July 1998, Rif, Snæfellsnessýsla, Iceland. © Gunnlaugur Þráinsson

12 July 1998, Fuglavík á Miðnesi, Gullbringusýsla, Iceland. © Jóhann Óli Hilmarsson

SPECIES ACCOUNTS

July 2003, Lough Errul, Cape Clear Island, Cork, Ireland. © Paul and Andrea Kelly

October 2004, Kenfig Pool, Glamorgan, Britain. © Alan Tate

- 29 October to 14 November 2003, Kenfig Pool, Glamorgan, male
- 09 and 13 March 2004, Llanilid, Glamorgan, male
- 13 October to 26 December 2004, Kenfig Pool, Glamorgan, male

3
- 20 September 2003 to 15 April 2004, Loch Tangasdale, Barra, Outer Hebrides, first-winter female [Fraser *et al.* 2007a; Rogers & Rarities Committee 2004; Scott 2004]
- 07–08 November 2004, Loch an Duin, Barra, Outer Hebrides, female

Iceland (2)

1 • 15 June to 10 July 1998, Rif, Snæfellsnessýsla, male [Kolbeinsson 2005; Kolbeinsson *et al.* 2001]

2 • 11–12 July 1998, Fuglavík á Miðnesi, Gullbringusýsla, adult male [Kolbeinsson 2005; Kolbeinsson *et al.* 2001]

Ireland (1)

1 • 12–15 July 2003, Lough Errul, Cape Clear Island, Cork, male [Milne 2005]

27

Spectacled Eider *Somateria fischeri*

Breeds on coasts in western and northern Alaska and northern Siberia from Lena delta eastward (del Hoyo *et al.* 1992; Madge & Burn 1988). Its winter quarters were discovered only recently (in 1995) in the Bering Sea, where 155 000 birds were counted (Balogh 1997; van den Berg 2004b).

A Norwegian record in May 1970 (cf. Haftorn 1971) was not accepted after review (Mjøs 2002). The record of four (at least two males) at Pechenga, Murmansk, Russia, in March 1938 (Meinertzhagen 1938) should be considered doubtful: although the published description seems convincing, Meinertzhagen's credibility has been questioned and, therefore, most (if not all) of his records are not accepted onto the British list (cf. Knox 1993b).

4 records (8 individuals)

Norway (3/5)

1 • 12 December 1933, Vardø, Finnmark, male, collected (BM: B.M. 9441) [Haftorn 1971; Johnsen 1937]
2 • 23–24 February 1988, Vardø, Finnmark, *three* (two males and one female) [Gustad 1995a]
3 • 15 June 1997, Mortensnes, Nesseby, Finnmark, adult male [Høyland *et al.* 2001]

Svalbard (1/3)

1 • 30 April 2002, north-west of Svalbard, at sea, *three* (two males and one female) [Bunes & Solbakken 2004]

12 December 1933, Vardø, Finnmark, Norway. © Ingvar Byrkjedal

White-winged Scoter *Melanitta deglandi*

June 1996, Smedaböle, Kemiö, Finland. © Sture Lindholm

4 December 1886, Crotoy, Somme, France. © Frédéric Jiguet

American White-winged Scoter *M.d. deglandi* breeds in North America from Alaska east to Hudson Bay and south to Manitoba and winters along both coasts south to Baja California and South Carolina; small numbers also winter in the Great Lakes region. Asian White-winged Scoter *M.d. stejnegeri* breeds in Asia from Yenisey basin east to Kamchatka and south to Mongolia and winters along Asian Pacific coast south to Korea, Japan and eastern China (del Hoyo *et al.* 1992; Madge & Burn 1988).

American and Asian White-winged Scoters have been recorded in the Western Palearctic: both in Iceland and Asian White-winged Scoter in Finland, France and Poland. It is difficult to assign the many sightings from Iceland to particular records and the sightings below might refer to more records than actually listed. The first Asian White-winged Scoter for Denmark occurred in October 2009 and again in March 2010 (van den Berg & Haas 2010c; Gantlett 2010). In Iceland, an American White-winged Scoter occurred in February–March 2010 (van den Berg & Haas 2010bc).

White-winged Scoter *Melanitta deglandi*

8 records (9 individuals)

Finland (1)
1 • 27 May to 08 June 1996, Smedaböle, Kemiö, adult male, trapped (27 May in a fishnet) (*M d stejnegeri*) [Lindroos 1997ab]

France (1)
1 • 04 December 1886, Crotoy, Somme, adult male, collected (MNHN: CG 1888-3821) (*M d stejnegeri*) [Jiguet & CAF 2007]

Iceland (5/6)
1 • 03 June 1993, Foss í Fossfirði, Vestur-Barðastrandarsýsla, male (*M d deglandi*) [Garðarsson 1997; Gunnlaugur Pétursson in litt; Kolbeinsson *et al.* 2003, 2006, 2008; Þráinsson *et al.* 1995; Þráinsson & Pétursson 1997]
• 23 June 2000, Reykjarfjörður í Suðurfjörðum, Vestur-Barðastrandarsýsla, adult male
• 17–27 June 2003, Reykjarfjörður í Suðurfjörðum, Vestur-Barðastrandarsýsla, male
• 27 May to 30 June 2005, Foss í Suðurfjörðum, Vestur-Barðastrandasýsla, male
• 17 July 2005, Reykjarfjörður í Suðurfjörðum, Vestur-Barðastrandarsýsla, adult male
2 • 04 June 1998, Þvottárskriður, Suður-Múlasýsla, *two* (males) (*M d*

23 June 2000, Reykjarfjörður í Suðurfjörðum, Vestur-Barðastrandarsýsla, Iceland. © Jóhann Óli Hilmarsson

12 June 2008, Bakkatjörn, Seltjarnarnes, Gullbringusýsla, Iceland. © Yann Kolbeinsson

24 April 2005, Hvalnesskriður við Álftafjörð, Austur-Skaftafellssýsla, Iceland. © Daniel Bergmann

SPECIES ACCOUNTS

deglandi) [Yann Kolbeinsson in litt; Kolbeinsson *et al.* 2001, 2004, 2008]
- 06 June 1998, Þvottárskriður, Suður-Múlasýsla, male
- 02 July 1998, Þvottárskriður, Suður-Múlasýsla, *two* (males)
- 12–17 July 2001, Þvottárskriður við Álftafjörð, Suður-Múlasýsla, adult male
- 24 April 2005, Hvalnesskriður við Álftafjörð, Austur-Skaftafellssýsla, male
- 30 April to 09 May and 02–07 July 2007, Þvottárskriður við Álftafjörð, Suður-Múlasýsla, male

3 • 06 April to 02 May 2003, Valþjófsstaðir í Núpasveit, Norður-Þingeyjarsýsla, adult male (*M d stejnegeri*) [Garner *et al.* 2004; Kolbeinsson *et al.* 2006]

4 • 20 May 2006, Kirkjuból í Skutulsfirði, Norður-Ísafjarðarsýsla, adult male (*M d deglandi*) [Þráinsson *et al.* 2009]

5 • 26 May to 12 June 2008, Bakkatjörn, Seltjarnarnes, Gullbringusýsla, male (*M d deglandi*) [Yann Kolbeinsson in litt]

Poland (1)

1 • 10 March 2007, Ptasi Raj, Gdańsk, Pomorskie, adult male (*M d stejnegeri*) [Komisja Faunistyczna 2008; Malczyk & Łukasik 2008]

1 May 2003, Valþjófsstaðir í Núpasveit, Norður-Þingeyjarsýsla, Iceland. © Dan Mangsbo

20 May 2006, Kirkjuból í Skutulsfirði, Norður-Ísafjarðarsýsla, Iceland. © Böðvar Þórisson

10 March 2007, Ptasi Raj, Gdańsk, Pomorskie, Poland. © D. Lukasik

Cape Teal *Anas capensis*

Breeds in Africa, from Sudan and Ethiopia southward to South Africa. Dispersive movements occur during dry season, especially to Lake Chad (del Hoyo *et al.* 1992; Madge & Burn 1988).

The two pairs in Libya were carrying nest material, prompting Cramp & Conder (1970) to believe they were breeding. All the records were during the dry season when birds move to areas with water.

6 records (10 individuals)

Israel (3/4)
1. • 26 June to 16 October 1978, Ashdod and Shifdan [Shirihai 1996]
2. • 15–19 May 1982, Shifdan [Shirihai 1996]
3. • 28 July 1984, Revadim reservoir, *two* [Shirihai 1996]

Libya (3/6)
1. • April 1961, 160 km north-east of Kufra oasis, Al Kufrah, remains found [Cramp & Conder 1970; Jany 1963]
2. • 03 April 1961, Kufra oasis, Al Kufrah [Cramp & Conder 1970; Jany 1963]
3. • at least 31 March to 05 April 1968, Kufra oasis, Al Kufrah, *two pairs* [Cramp & Conder 1970]

Red-billed Teal *Anas erythrorhyncha*

Breeds in Africa, from southern Sudan and Ethiopia south to South Africa; also Madagascar. Dispersive movements occur in dry years (del Hoyo *et al.* 1992; Madge & Burn 1988).

The Israeli record was believed to be a genuine vagrant by Shirihai (1996), an opinion still shared by the Israeli rarities committee despite the fact that it also occurs in captivity. It has also occasionally turned up in Europe.

1 record

Israel (1)
1. • 20 June to 12 July 1958, Ma'agan Mikhael [Shirihai 1996]

LOONS Gaviidae (n=1)

Pacific Loon *Gavia pacifica*

Breeds in northern North America. Winters off Pacific coast of North America (del Hoyo *et al.* 1992).

This loon has probably been overlooked in the past due to its similarity to Black-throated Loon *G. arctica*. In Britain, the birds in Pembrokeshire and Cornwall returned in 2009, while a new bird occurred in Gloucestershire in November 2009 (Hudson & Rarities Committee 2010; Millington 2009). The first for Spain occurred in December 2009 (Gantlett 2010; Velaco 2010). The first for the Channel Islands occurred from mid-December 2009 to January 2010 and was seen again in March 2010 (van den Berg & Haas 2010ac; Lawlor 2010). The first for Ireland was found in January 2010 and stayed into May 2010 (van den Berg & Haas 2010bc).

3 records

Britain (3)

1. • 12 January to 04 February 2007, Farnham Gravel Pits, North Yorkshire, juvenile [British Ornithologists' Union 2010; Hudson & Rarities Committee 2009; Mather 2010; Taylor *et al.* 2007]
2. • 02 February to 20 March 2007, Llys-y-Fran Reservoir, Pembrokeshire, juvenile [Astins & Brown 2007; British Ornithologists' Union 2010; Hudson & Rarities Committee 2009; Mullarney & Millington 2008]
   • 16 January to 11 February 2008, Llys-y-Fran Reservoir, Pembrokeshire, second-winter
3. • 17 February to 10 March 2007, Mount's Bay, Penzance, Cornwall, adult [Ahmad 2007; British Ornithologists' Union 2010; Hudson & Rarities Committee 2009]
   • 23–29 November 2007, Marazion, Cornwall, adult

17 February 2007, Mount's Bay, Penzance, Cornwall, Britain. © Lee Fuller

Extremely rare birds in the Western Palearctic

Pacific Loon *Gavia pacifica*

January 2007, Farnham Gravel Pits, North Yorkshire, Britain. © Steve Young

February 2008, Llys-y-fran Reservoir, Pembrokeshire. © Richard Stonier

ALBATROSSES
Diomedeidae (n=3)

Atlantic Yellow-nosed Albatross
Thalassarche chlororhynchos

Breeds in South Atlantic in Tristan da Cunha group and on Gough Island. Some birds move east off west coast of South Africa and west off east coast of South America. Vagrants have been found off the east coast of North America (Onley & Scofield 2007). It was also recorded in Greenland in August 1944 and August 1963 (Boertmann 1994).

The second record for Norway occurred in June and July 2007, but has not yet been submitted to the Norwegian rarities committee (cf. Gantlett & Plym 2007; Tor A Olsen in litt). A record of a skeleton obtained in Iceland in c 1844 (Cramp & Simmons 1977) is not accepted. Furthermore, a sighting off Cornwall, Britain, in April 1985 has been withdrawn by the observer because of the time the British rarities committee and BOU records committee took to consider this record (cf. British Ornithologists' Union 1994a; Curtis 1993).

3 records

Britain (1)
1 • 29 June 2007, Brean Down, Somerset, immature, taken into care and released (30 June) [British Ornithologists' Union 2010; Gantlett & Pym 2007; Hudson & Rarities Committee 2009; Rowlands *et al.* 2010] [1]
• 02–03 July 2007, Manton, Messingham, Lincolnshire, immature

Norway (1)
1 • 13 April 1994, 64° 28 N 09° 45 E, 20 nautical miles northwest off Halten, Sør-Trøndelag, at sea, adult [Gustad 1995a]

Sweden (1)
1 • 08 July 2007, Domsten, Landskrona-Barsebäck-Lomma, Skåne, immature [Hellström & Strid 2008; Gantlett & Pym 2007] [1]
• 08 July 2007, Landskrona-Barsebäck-Lomma, Skåne, immature
• 08 July 2007, Malmö, Skåne, immature

[1] Same individual

Extremely rare birds in the Western Palearctic

Atlantic Yellow-nosed Albatross *Thalassarche chlororhynchos*

3 July 2007, Manton, Messingham, Lincolnshire, Britain. © Paul Condon

8 July 2007, Malmö, Skåne, Sweden. © Kristian Stahl

13 April 1994, off Halten, Sør-Trøndelag, at sea, Norway. © Jostein Sørgård

Shy Albatross *Thalassarche cauta*

Breeds on Albatross Island in Bass Strait, off Tasmania. Dispersive movements occur in southern oceans (Onley & Scofield 2007).

The sole Western Palearctic record concerned a bird first seen at Eilat, Israel, and also off Taba, Egypt. It was later found dead and identified as a subadult male. There have been several records off Somalia and Kenya, suggesting that the Western Palearctic bird travelled through the Arabian Sea into the Red Sea (Baker 1989; Gichuki & Pearson 1987; Meeth & Meeth 1988).

1 record

Egypt (1)

1 • between 20 February and 07 March 1981, off Taba, subadult male [Goodman & Meininger 1989; Goodman & Storer 1987] [1]

Israel (1)

1 • 20–26 February and 02 March 1981, North Beach, Eilat, subadult male, collected (07 March) (TAU: 9659) [Goodman & Storer 1987; Shirihai 1996] [1]

[1] Same individual

7 March 1981, North Beach, Eilat, Israel. © Oz Horine

Tristan Albatross *Diomedea dabbenena*

Breeds on Gough Island and islands in Tristan da Cunha group. Dispersive movements occur in South Atlantic between 23° S and 42° S (Onley & Scofield 2007).

Soldaat *et al.* (2009) listed 19 records and reports of *Diomedea* albatrosses in the Western Palearctic. Of these, the specimen from Sicily was identified as a Tristan Albatross and is currently accepted on the Italian list (Giancarlo Fracasso in litt). There have been at least three additional reports in the 19th century (two in Belgium and one in France), which have been treated as 'dubious' records, most likely concerning imported birds trapped by sailors (cf. Bourne 1967). A report of an immature c 80 km off Portugal in October 1963 (Bourne 1966) is now under review by the Portuguese rarities committee (João Jara in litt).

1 record

Italy (1)

1 • 04 October 1957, Termini Imerese, Palermo, Sicily, collected (MSNT: 2537 ex 4300), immature male [Corso 2009; Haas 2009; Orlando 1958]

4 October 1957, Termini Imerese, Palermo, Sicily, Italy. © Fabio Lo Valvo

PETRELS
Procellariidae (n=11)

Southern Giant Petrel *Macronectes giganteus*

Breeds on islands in southern oceans and on Antarctic peninsula. Dispersive movements occur anywhere in southern oceans (Onley & Scofield 2007).

The Italian bird was seen from a ferry between Greece and Ancona, Italy (Frédéric Jiguet in litt). A bird off Ouessant, Finistère, France, in November 1967, was accepted as a giant petrel *Macronectes* (cf. Meeth 1969; Dubois *et al.* 2008).

1 record

Italy (1)
1 • 02 September 1991, Adriatic Sea [Brichetti *et al.* 1995]

Cape Petrel *Daption capense*

Breeds on islands in southern oceans and on Antarctic peninsula. Dispersive movements occur anywhere in southern oceans (Onley & Scofield 2007).

Haas & Crochet (2009) reviewed all Western Palearctic records, including a total of 10 records and reports from Britain, France, Ireland, Italy and the Netherlands. The only category A record remains the Gibraltar record. However, Fracasso *et al.* (2009) still list an Italian record in category A, which is listed below as well. In May 2010, a bird was photographed off Norway (van den Berg & Haas 2010c).

2 records

Gibraltar (1)
1 • 20 June 1979, Europa Point [Haas & Crochet 2009; Holliday 1990]

Italy (1)
1 • September 1964, off Sciacca, Agrigento, Sicily, at sea, immature, collected (private collection) [Haas & Crochet 2009; Massa 1974]

September 1964, off Sciacca, Agrigento, Sicily, at sea, Italy. © Bruno Massa

Trindade Petrel *Pterodroma arminjoniana*

Breeds on Round Island near Mauritius in Indian Ocean and on Trindade Island and Martin Vaz Rocks in South Atlantic. Regularly seen off North Carolina in North Atlantic (Onley & Scofield 2007).

There have been an additional six records in the Azores: one photographed in May 2006 and singles seen in October 2006, October 2007, August 2009, April 2010 and May 2010 (van den Berg & Haas 2010c; Gantlett 2010; Pinhuinhas 2006). The six recent records in the Azores suggest that this species may occur more commonly in the North Atlantic than once previously thought.

2 records

Azores (1)
1 • 18 July 1997, 10 nautical miles south of Pico, at sea, dark morph [Costa *et al.* 2000; Dubois & Seitre 1997]

Cape Verde (1)
1 • 30 September 2008, 5 nautical miles off southern tip of Brava, intermediate morph [Hazevoet 2010]

30 September 2008, off Brava, Cape Verde.
© Lindblad Expeditions

17 May 2006, off Faial, Azores.
© Manuel Fernandes

Extremely rare birds in the Western Palearctic

Soft-plumaged Petrel *Pterodroma mollis*

Breeds on islands in south-western Pacific, South Atlantic and southern Indian Oceans. Disperses north as far as 35° S (Onley & Scofield 2007).

This two-nation record is the only one for the Western Palearctic. At-sea identification of *Pterodroma* petrels is notoriously difficult. Presumably, most records in north-western Europe (especially Britain and Ireland) relate to Fea's Petrel *P. feae*. A Soft-plumaged Petrel was photographed in Norway in June 2009 (Catley 2009; Gantlett 2010).

1 record

Israel (1)
1 • 25 March 1997, North Beach, Eilat [Israeli Rarities and Distribution Committee 2002; Sapir & Israeli Rarities and Distribution Committee 2007; Shirihai 1999] [1]

Jordan (1)
1 • 25 March 1997, Aqaba [Andrews *et al.* 1999] [1]

[1] Same individual

Bermuda Petrel *Pterodroma cahow*

Breeds on rocky islets in Castle Harbour, Bermuda. It is thought to move north into the North Atlantic but dispersive movements are still poorly known. There are confirmed records off North Carolina (Onley & Scofield 2007).

A remarkable occurrence of this Bermudan endemic, which was trapped on four occasions at the same site in 2002–06.

1 record

Azores (1)
1 • 17 November 2002, Ilhéu da Vila, Santa Maria, adult, trapped [Joël Bried in litt; Bried 2003; Elias *et al.* 2004; Jara *et al.* 2008]
  • 19 and 21 November 2003, Ilhéu da Vila, Santa Maria, adult, trapped
  • 12 / 13 December 2006, Ilhéu da Vila, Santa Maria, adult, trapped

17 November 2002,
Ilhéu da Vila,
Santa Maria, Azores.
© Joël Bried

Black-capped Petrel *Pterodroma hasitata*

Breeds on islands in Caribbean. Dispersive movements occur into north-western North Atlantic (Onley & Scofield 2007).

A record off Scotland, Britain, in February 1980 (Dannenberg 1983) was not accepted by the British rarities committee (cf. Rogers & Rarities Committee 1986). A French report of a bird collected in Pas-de-Calais in the 19th century was not accepted because the specimen is missing (cf. CAF 2006). The second record for the Azores was in May 2009 (Gantlett 2010). In May 2010, a bird was seen off Madeira (van den Berg & Haas 2010c).

4 records

Azores (1)
1 • 26 May 2007, 16 km south-east of Graciosa, at sea [João Jara in litt]

Britain (2)
1 • March / April 1850, Southacre, Swaffham, Norfolk, collected (CM: Acc. No. 108.949) [Bourne 1967; Newton 1852]
2 • 16 December 1984, Barmston, East Yorkshire, first-year female, found dead (private collection) [British Ornithologists' Union 1991a, 2006a; Mather & Curtis 1987; Rogers & Rarities Committee 1986]

Spain (1)
1 • 30 April 2002, 45° 01 N 12° 16 W, 180 nautical miles northwest off Cabo Finisterre, at sea [de Juana 2006; Howell 2002; de Juana & Comité de Rarezas de la SEO 2004]

26 May 2007, off Graciosa, at sea, Azores.
© Killian Mullarney

March or April 1850, Southacre, Swaffham, Norfolk, Britain. © Phil Palmer

Atlantic Petrel *Pterodroma incerta*

Breeds in South Atlantic in Tristan da Cunha group and on Gough Island. Some disperse east off west coast of South Africa and west off east coast of South America (Onley & Scofield 2007).

A Slovakian specimen (Bourne 1992) was not accepted since it is missing and could not be examined (cf. Trnka & Matousek 1996).

2 records

Israel (2)
1. • 31 May 1982, North Beach, Eilat [Shirihai 1996, 1999] [1]
2. • 18–24 April 1989, North Beach, Eilat [van der Schot 1989; Shirihai 1996, 1999] [2]

Jordan (2)
1. • 31 May 1982, Aqaba [Andrews *et al.* 1999] [1]
2. • 18–24 April 1989, Aqaba [Andrews *et al.* 1999] [2]

[1] Same individual
[2] Same individual

18 April 1989, North Beach, Eilat, Israel.
© Stefan Golaszewsky

Streaked Shearwater *Calonectris leucomelas*

Breeds in north-western Pacific Ocean on islands off Japan, eastern China, Korea and south-eastern Russia. Dispersive movements occur south into Australian waters and west into Indian Ocean (Onley & Scofield 2007). It has also been recorded as a vagrant in Oman (Eriksen *et al.* 2003).

The records in 1992–93 concerned birds that lingered for a certain length of time.

3 records (4–6 individuals)

Israel (3/4–6)

1 • 27 June to 18 September 1992, North Beach, Eilat, *two or three* [Morgan & Shirihai 1992; Shirihai 1996, 1999] [1]

2 • early May to summer of 1993, North Beach, Eilat [Shirihai 1996, 1999] [2]

3 • 02 February 2003, North Beach, Eilat, *one or two* [Shochat *et al.* 2004]

Jordan (2/3–4)

1 • 27 June to 18 September 1992, Aqaba, *two or three* [Andrews *et al.* 1999] [1]

2 • early May to summer of 1993, Aqaba [Andrews *et al.* 1999] [2]

[1] Same individual
[2] Same individual

June 1992, North Beach, Eilat, Israel.
© Hadoram Shirihai

Flesh-footed Shearwater *Puffinus carneipes*

Breeds on islands in Pacific and Indian Oceans. Most migrate north into Northern Hemisphere (e.g. off Japan and Gulf of Alaska) (Onley & Scofield 2007). A fairly common summer visitor off southern Oman (Eriksen *et al.* 2003) and has also been recorded as a vagrant in UAE (Pedersen & Aspinall 2010).

The sole record concerned a bird probably originating from the Indian Ocean population.

1 record

Israel (1)
1 • 15 August 1980, North Beach, Eilat [Shirihai 1996, 1999] [1]

Jordan (1)
1 • 15 August 1980, Aqaba [Andrews *et al.* 1999] [1]

[1] Same individual

Wedge-tailed Shearwater *Puffinus pacificus*

Breeds on islands throughout tropical Pacific and Indian Oceans. Dispersive movements are not fully understood but they may reach central-northern and eastern Pacific (Onley & Scofield 2007).

A report at Quseir, Egypt, in November 1983 (Bezzel 1987) was not accepted by Goodman & Meininger (1989).

1 record

Egypt (1)
1 • 10 March 1988, off Port Said, Red Sea, at sea [Everett 1992]

SPECIES ACCOUNTS

Tropical Shearwater *Puffinus bailloni*

Breeds in Mascarene Islands (Réunion and Europa Island) in Indian Ocean. Probably sedentary but birds have occurred outside known range (Onley & Scofield 2007).

Shirihai & Sinclair (1994) mentioned the record of an unidentified small shearwater off Eilat believed at first to be a new species *P. atrodorsalis* (Shirihai 1999; Shirihai *et al.* 1995), but now known to refer to Tropical Shearwater (cf. Sapir *et al.* 2007). The May 1999 record was also a sight record, but the bird was not photographed.

2 records

Israel (2)
1. 18–21 June 1992, Gulf of Aqaba, Eilat [Sapir & Israeli Rarities and Distribution Committee 2007; Shirihai & Sinclair 1994]
2. 15 May 1999, Gulf of Aqaba, Eilat [Sapir & Israeli Rarities and Distribution Committee 2007]

June 1992, Gulf of Aqaba, Eilat, Israel. © Hadoram Shirihai

47

TROPICBIRDS
Phaethontidae (n=1)

White-tailed Tropicbird *Phaethon lepturus*

Breeds on islands in Pacific, Indian and Atlantic Oceans; in eastern and central Atlantic, on Ascension and islands in Gulf of Guinea; in western Atlantic, on Bermuda, Bahamas, islands in Caribbean and on Fernando de Noronha (del Hoyo *et al.* 1992).

The Cape Verde bird was seen while searching for Magnificent Frigatebird *Fregata magnificens* at its only breeding site in the Western Palearctic. It was seen at long range, but well enough to identify it. Hazevoet (1995) mentions two records close to Cape Verde, but outside Western Palearctic waters – one at 17° 05 N 18° 26 W on 8 July 1975 and one 166 km south of Santiago on 4 December 1988 – and so the occurrence of White-tailed Tropicbird in Cape Verde is to be expected. In October 2007, an immature was photographed 200–300 nautical miles west of the Azores, outside Western Palearctic waters (Gantlett 2008).

1 record

Cape Verde (1)
1 • 20 February 1999, Ilhéu de Curral Velho, Boavista, adult [Dufourny 1999; Hazevoet 1999]

SPECIES ACCOUNTS

GANNETS Sulidae (n=2)

Red-footed Booby *Sula sula*

Breeds on islands in South Atlantic, Pacific and Indian Oceans and in Caribbean Sea (del Hoyo *et al.* 1992). It has also been recorded as a vagrant in Oman and UAE (Eriksen *et al.* 2003, 2010; Pedersen 2010).

A report at Mølen, Larvik, Vestfold, Norway, on 29 June 1985 was included on the Norwegian list for many years (Bentz 1988) but was removed in 2002 due to insufficient documentation (cf. Mjøs 2002). Recently, two records have been made public of single birds photographed between Cape Verde and the Canary Islands in October 2004 and July 2005 (José Luis Copete *in litt*). Hazevoet (2010) mentioned two additional records from Cape Verde in October–November 2009.

4 records

Cape Verde (4)
1. • 17 April 1977, c 17° N 23° W, just north of Sal, at sea, came on board *MV Causeway* [Hazevoet 2010; Nuovo 2008]
2. • 24 August 1986, Ilhéu de Cima, Ilhéus do Rombo, adult, white-tailed brown morph [den Hartog 1987, Hazevoet 1995]
3. • 21 July 2005, Ponta do Sol, Santa Antão [Hazevoet 2010]
4. • 21 October 2007, c 350 km north of Cape Verde, at sea, subadult [Gantlett 2008; Hazevoet 2010]

LEFT: 17 April 1977, off Sal, at sea, Cape Verde. © Ernest Glyndwr Lewis

BELOW: 21 October 2007, c. 350 km north of Cape Verde, at sea, Cape Verde. © Bud Lehnhausen

Masked Booby *Sula dactylatra*

Breeds on islands in South Atlantic, Pacific and Indian Oceans and in Caribbean Sea (del Hoyo *et al.* 1992) and is a common visitor off southern Oman (Eriksen *et al.* 2003). It has also been recorded as a vagrant in UAE (Pedersen & Aspinall 2010).

The Spanish records from 1985 might refer to the same bird.

16 July 2004, Rishon Letzion, Israel. © Itzik Amir

7 records

Azores (1)
1. • 01 August 2008, four miles off Cedros, Faial, adult [João Jara in litt]

Cape Verde (1)
1. • 2003–05, Ilhéu de Curral Velho, Boavista, adult [Hazevoet 2010]

France (1)
1. • 03 September 2003, 46° 26 N 4° 04 W, Rochebonne, c 150 km west of Croix-de-Vie, Vendée, at sea, subadult, following Portsmouth–Bilbao ferry [Frémont & CHN 2005; Harrop 2003ab; Jiguet & CAF 2004] [1]

Israel (1)
1. • 16 July 2004, Rishon Letzion, adult [Mizrachi *et al.* 2007; Sapir & Israeli Rarities and Distribution Committee 2007]

Morocco (1)
1. • 16 January 2006, between Tantan Plage and Oued Chebeika, Tarfaya, subadult [Bergier *et al.* 2008]

Spain (3)
1. • 10 October 1985, Puerto Sotogrande, Cadiz, adult [de Juana 2006; de Juana & Comité Ibérico de Rarezas de la SEO 1988]
2. • 14 December 1985, Torremolinos, Málaga, adult [de Juana 2006; de Juana & Comité Ibérico de Rarezas de la SEO 1988]
3. • 04 September 2003, Bilbao harbour, Bay of Biscay, at sea, subadult, following Portsmouth–Bilbao ferry [Harrop 2003a; de Juana 2006; de Juana & Comité de Rarezas de la SEO 2005] [1]

[1] Same individual

SPECIES ACCOUNTS

1 August 2008, four miles off Cedros, Faial, Azores. © Lisa Steiner / Whale Watch Azores

3 September 2003, Croix-de-Vie, Vendée, at sea, France. © Hugh Harrop

22 April 2005, Ilhéu de Curral Velho, Boavista, Cape Verde. © Pedro López Suárez

FRIGATEBIRDS
Fregatidae (n=2)

Ascension Frigatebird *Fregata aquila*

Endemic to Ascension island in South Atlantic. Disperses to surrounding waters (del Hoyo *et al.* 1992).

Anonymous (1954) detailed a record of a Magnificent Frigatebird *F. magnificens* found moribund in July 1953. This record has long been on the British list as the sole record but Walbridge *et al.* (2003) re-identified it as an immature female Ascension Frigatebird.

1 record

Britain (1)

1 • 10 July 1953, Tiree, Argyll, immature female, taken into care and died (NMS: Z.1953.16) [British Ornithologists' Union 1956, 2004; Rogers & Rarities Committee 2003; Walbridge *et al.* 2003]

10 July 1953, Tiree, Argyll, Britain. © National Museums Scotland

SPECIES ACCOUNTS

Lesser Frigatebird *Fregata ariel*

Breeds on islands in Indian and Pacific Oceans and on Trindade off Brazil in South Atlantic (del Hoyo *et al.* 1992). It has also been recorded as a vagrant in Oman (Eriksen *et al.* 2003).

The three records probably involve birds from islands in the western Indian Ocean.

10 April 2008, Zour Port, Kuwait. © Lee Gregory

3 records

Israel (2)
1. • 01 December 1997, North Beach, Eilat, immature male [Riddington & Reid 2000; Sapir & Israeli Rarities and Distribution Committee 2007; Smith *et al.* 2001] [1]
2. • 06 May 1999, North Beach, Eilat [Sapir & Israeli Rarities and Distribution Committee 2007; Smith *et al.* 2002; van Welie 2000]

Jordan (1)
1. • 01 December 1997, Aqaba, immature male [Andrews *et al.* 1999] [1]

Kuwait (1)
1. • 10 April 2008, Zour Port, immature [Lansdell *et al.* 2008; Mike Pope in litt]

[1] Same individual

Extremely rare birds in the Western Palearctic

HERONS Ardeidae (n=8)

Least Bittern *Ixobrychus exilis*

Breeds in eastern and central North America, from southern Canada to Gulf of Mexico and Caribbean. Winters in southern North America, Caribbean and northern South America (Cramp & Simmons 1977; del Hoyo *et al.* 1992).

A British report of a bird collected at York in the autumn of 1852 was not accepted by the British rarities committee but was mentioned by Evans (1994) as a genuine record.

5 November 2001, Angra do Heroísmo, Terceira, Azores. © Filipe Barata

28 September 2007, Vila do Porto, Santa Maria, Azores. © Carlos Pereira

SPECIES ACCOUNTS

27 November 1951, Ponta Delgada, São Miguel, Azores. © Museu Carlos Machado / Staffan Rodebrand

8 October 1964, Ponta Delgada, São Miguel, Azores. © Museu Carlos Machado / Staffan Rodebrand

11 September 1952, Povoacao, São Miguel, Azores. © Museu Carlos Machado / Staffan Rodebrand

RIGHT: 17 September 1970, Búastaðir á Heimaey, Vestmannaeyjar, Iceland. © Yann Kolbeinsson

8 records

Azores (7)

1. • sine dato, Santa Maria, collected (specimen location unknown) [Le Grand 1983]
2. • 07 September 1951, Terceira, female, collected (specimen location unknown) [Bannerman & Bannerman 1966]
3. • 27 November 1951, Ponta Delgada, São Miguel, female, collected (MCM: MCM1371) [Bannerman & Bannerman 1966]
4. • 11 September 1952, Provocão, São Miguel, female, collected (specimen location unknown) [Bannerman & Bannerman 1966]
5. • 08 October 1964, Ponta Delgada, São Miguel, male, collected (MCM: MCM1632) [Bannerman & Bannerman 1966]
6. • 05 November 2001, Angra do Heroísmo, Terceira, first-winter female, taken into care and released [João Jara in litt]
7. • 28 September 2007, Vila do Porto, Santa Maria, trapped, juvenile [Jara et al. 2008]

Iceland (1)

1. • 17 September 1970, Búastaðir á Heimaey, Vestmannaeyjar, collected (IMNH: RM262) [Pétursson & Þráinsson 1999]

Von Schrenck's Bittern *Ixobrychus eurhythmus*

Breeds from south-eastern Siberia, Manchuria and Japan south to eastern China. Winters from southern China, Indochina and Malay peninsula to Greater Sundas, Sulawesi and Philippines (del Hoyo *et al.* 1992).

Haas *et al.* (in prep.) reviewed its status in the Western Palearctic and concluded that the Italian record is genuine. A report of a female shot in Germany between 1895 and 1897 was not accepted because of doubts about its location and the date of collection.

1 record

Italy (1)
1 • 12 November 1912, Bra, Cuneo, Piemonte, first-winter female, collected (MRSN: MZUT-12593) [Boano & Mingozzi 1986; Salvadori 1912–13]

12 November 1912, Bra, Cuneo, Piemonte, Italy. © Giovanni Boano

SPECIES ACCOUNTS

Dwarf Bittern *Ixobrychus sturmii*

Breeds in sub-Saharan Africa (del Hoyo *et al.* 1992).

Two additional reports (Velmala *et al.* 2002) have not (yet) been accepted. A report of two birds shot in the 19th century in France is not accepted because the specimens could not be traced (CAF 2006).

3 records

Canary Islands (3)

1 • October 1886, La Laguna, Tenerife, adult male, collected (ICPL) [Díes *et al.* 2007]
2 • 21–30 January 2000, Aldea Blanca ponds, Gran Canaria, first-winter [de Juana 2006; de Juana & Comité de Rarezas de la SEO 2002]
3 • 23 August 2002 to 10 May 2003, Erjos ponds, Tenerife, adult male [de Juana 2006; de Juana & Comité de Rarezas de la SEO 2004, 2005; Velmala *et al.* 2002]

October 1886, La Laguna, Tenerife, Canary Islands. © Juan Antonio Lorenzo

August 2002, Erjos ponds, Tenerife, Canary Islands. © Ludovic Scalabre

Indian Pond Heron *Ardeola grayii*

Breeds in northern Persian Gulf east through India and Sri Lanka to Burma, also Laccadives, Maldives, Andamans and Nicobars. Largely sedentary but some local movements related to the occurrence of floods and droughts (del Hoyo *et al.* 1992).

Its status in Kuwait is unclear (cf. Cramp & Simmons 1977) and, for instance, it was not mentioned in Gregory (2005). Thus, the only record is from Egypt. However, a bird was photographed in Kuwait in November 2009 (Gantlett 2010; Mike Pope in litt). A report in Belgium in July 1988 was not accepted as a genuine vagrant (cf. Boesman 1990; Claeys *et al.* 1989).

1 record

Egypt (1)
1 • 24 April 2004, Crocodile Island [Jiguet 2006]

24 April 2004, Crocodile Island, Egypt. © Frédéric Jiguet

SPECIES ACCOUNTS

Little Blue Heron *Egretta caerulea*

Breeds on coasts of southern North America, Central America and northern half of South America (del Hoyo *et al.* 1992). It was also recorded in Greenland in September 1942 (Boertmann 1994).

The first record is a ringing recovery (cf. Dennis 1981).

4 records

Azores (3)

1 • 28 November 1964, Fazendas de Santa Cruz, Flores, first-winter (ringed as nestling in New Jersey, USA, in June 1964) [Dennis 1981]
2 • 07–09 October 1997, Fajã dos Cubres, São Jorge, first-winter [Costa *et al.* 2000]
3 • 04–07 October 1998, Madalena, Pico, first-winter [Costa *et al.* 2003]

Ireland (1)

1 • 24 September to 22 October 2008, Letterfrack, Galway, juvenile [Paul Milne in litt; Breen 2008]

ABOVE: October 1998, Madalena, Pico, Azores. © Gonçalo Elias

BELOW: 5 October 2008, Letterfrack, Galway, Ireland. © Chris Batty

Tricolored Heron *Egretta tricolor*

Breeds on coasts of southern North America, Central America and northern South America. North American populations winter south (del Hoyo *et al.* 1992).

The bird on São Miguel, Azores, in October 2007 might be the same as the one in the Canary Islands from November 2007 to July 2008.

3 records

Azores (2)
1. • 22–24 October 1985, Lajes do Pico, Pico, immature [Parrot *et al.* 1987]
2. • 02 October 2007, Praia de Água de Alto, São Miguel, first-winter [João Jara *in litt*]

Canary Islands (1)
1. • 15 November to 01 December 2007, Playa de las Canteras, Gran Canaria, first-winter [Bonser 2008; Díes *et al.* 2009; Gantlett 2009]
   • 08 December 2007 to 04 July 2008, Playa de las Americas, Tenerife, first-winter

2 October 2007, Praia de Água de Alto, São Miguel, Azores. © Silke Sottorf

October 1985, Lajes do Pico, Pico, Azores. © John Parrot

13 February 2008, Playa de las Americas, Tenerife, Canary Islands. © Chris Batty

Black Heron *Egretta ardesiaca*

Breeds in sub-Saharan Africa, except Congo basin and area around Kalahari; also on Madagascar (del Hoyo *et al.* 1992).

The third for Cape Verde was photographed in February 2010 (Kees Hazevoet in litt). A report of two birds at Aswan, Egypt, in August 1980 (cf. Dekker 1981) was not accepted by Goodman & Meininger (1989).

3 records

Cape Verde (2)
1 • 06 February and 07 March 1985, Ilhéu de Curral Velho, Boavista [Hazevoet 1995]
2 • 06 March 2007, Razo [Gantlett 2008; Hazevoet 2010]

Israel (1)
1 • 19–20 October 1982, Eilat [Shirihai 1996, 1999]

Black-headed Heron *Ardea melanocephala*

Breeds in Africa, from Senegal east to Ethiopia and south to South Africa. Largely sedentary but some birds move northward during dry season (del Hoyo *et al.* 1992). It has also been recorded as a vagrant in Oman (Eriksen *et al.* 2003, 2010).

A French report of a bird shot in Var in 1845 was not accepted because its origin was considered doubtful (CAF 2006). The first for Cape Verde occurred on Santiago in March–April 2009 (Gantlett 2010; Hazevoet 2010).

3 records

France (2)
1 • end of 19th century, Saintes-Maries-de-la-Mer, Bouches-du-Rhône, collected (specimen location unknown) [Dubois *et al.* 2008]
2 • 29 November 1971, Camargue, Bouches-du-Rhône, adult [Dubois *et al.* 2008; Hovette 1972; Hovette & Kowalski 1972]

Israel (1)
1 • 19 October to 15 December 1987, Eilat, immature [Shirihai 1996, 1999] [1]

Jordan (1)
1 • 19 October to 15 December 1987, Aqaba, immature [Andrews *et al.* 1999] [1]

[1] Same individual

STORKS Ciconiidae (n=1)

Marabou Stork *Leptoptilos crumeniferus*

Breeds in sub-Saharan Africa, from Senegal to Sudan and south to southern Africa. Some dispersive movements occur (Cramp & Simmons 1977; del Hoyo *et al.* 1992).

Records in Spain and Portugal are placed in category D (cf. de Juana 2006).

3 records

Israel (2)
1. • May 1951, Hula valley, collected (specimen location unknown) [Shirihai 1996]
2. • March–May 1957, Kibbutz Dgania, Lake Tiberias, Jordan valley, *two*, one collected (31 March) (BGM) [Nir Sapir in litt; Shirihai 1996]

Mauritania (1)
1. • 29 March 1966, Arel, Banc d'Arguin [Isenmann 2007; Mahé 1985; de Naurois 1969]

31 March 1957, Kibbutz Dgania, Lake Tiberias, Jordan valley, Israel. © Adi Teitler

HAWKS
Accipitridae (n=4)

Swallow-tailed Kite *Elanoides forficatus*

Breeds in North America, from Florida along coast west to Texas; also in Central and northern South America. Winters in northern South America (del Hoyo *et al.* 1994; Meyer 1995).

A record from Fuerteventura, Canary Islands, in March 1993 was placed in category D (cf. Müller & Lippert 1998; de Juana 2006; de Juana & Comité de Rarezas de la SEO 1998). A report from the Azores in March 2005 is now under review (João Jara in litt).

1 record

Azores (1)
1 • 24 August to 07 September 2008, Vigia das Feteiras, São Miguel [João Jara in litt]

28 August 2008, Vigia das Feteiras, São Miguel, Azores. © Duarte Araújo

African Fish Eagle *Haliaeetus vocifer*

Breeds in Africa, from Senegal to Ethiopia and south to South Africa (Cramp & Simmons 1980; del Hoyo *et al.* 1994).

A report of three birds in Sinai in November 1967 (Cramp & Simmons 1980) was not accepted by Goodman & Meininger (1989). A report of a juvenile at Garf Hussein, Lake Nasser, Egypt, in October 1997 (Gantlett 1999) was not accepted because details are lacking.

1 record

Egypt (1)

1 • 01 November 1947, Abu Handel, south of Aswan, adult, collected (GZM) [Goodman & Meininger 1989]

Bald Eagle *Haliaeetus leucocephalus*

Breeds in North America, south to north-western Mexico. Winters in central North America and on coasts (del Hoyo *et al.* 1994).

In Britain, an adult seen in Gwynedd in October 1978 was not accepted as a genuine vagrant (cf. British Ornithologists' Union 1980, 2009; Evans 1994; Rogers & Rarities Committee 1980). Furthermore, a specimen allegedly trapped in January 1865 in North Yorkshire was not accepted because of a possible mix-up of specimens (Clancey 1950; Andrew Harrap in litt).

2 records

Ireland (1)

1 • 15 November 1987, Ballymaceligot, Castleisland, Kerry, juvenile, taken into care, released in the USA in December [Grant 1987; Smiddy & O'Sullivan 1996]

Northern Ireland (1)

1 • 11 January 1973, Garrison, Fermanagh, juvenile, collected (UM: Lg7116) [British Ornithologists' Union 1999; Smiddy & O'Sullivan 1996]

15 November 1987, Ballymacelligot, near Castleisland, Kerry, Ireland. © Richard Mills

11 January 1973, near Garrison, Fermanagh, Northern Ireland. © The Ulster Museum

Hooded Vulture *Necrosyrtes monachus*

Breeds in Africa, from southern Mauritania east to Ethiopia and south to Namibia and South Africa. Mainly resident but some dispersive movements noted in June–September (del Hoyo *et al.* 1994).

A report in Spain in February 2003 was not accepted as a genuine vagrant (de Juana & Comité de Rarezas de la SEO 2005). Thévenot *et al.* (2003) mention a possible record in Morocco in April 1989.

4 records (5 individuals)

Mauritania (3)
1 • December 1960, Nouadhibou [Heim de Balsac & Mayaud 1962; Isenmann 2007]
2 • December 1960, Banc d'Arguin [Heim de Balsac & Mayaud 1962; Isenmann 2007]
3 • 08–09 April 2001, Cape Timiris, Banc d'Arguin [Isenmann 2007]

Western Sahara (1/2)
1 • 07 June 1955, Sbayera, *two* [Valverde 1957]

65

Extremely rare birds in the Western Palearctic

FALCONS
Falconidae (n=2)

American Kestrel *Falco sparverius*

Breeds across most of North America throughout Central and South America. Northernmost populations are migratory and move south to winter in southern North America (del Hoyo *et al.* 1994).

There has been no genuine record in the Western Palearctic since 1980. An adult male was trapped in Estonia in June 1963 and was considered to be an escape (Margus Ots in litt). In Britain, a very tame male frequented Dungeness, Kent, from June 1966 to April 1967 and was also considered an escape (cf. Evans 1994). One in Switzerland in December 1990 was considered an escape because of its plumage and tame behaviour (Undeland & Leuzinger 1992). In July–August 2003, three males were present in Germany and in September 2004 possibly the same male was reported in Britain (Anonymous 2003; Deutsche Seltenheitenkommission 2008). A ringed female occurred in Britain in October 2005 (Anonymous 2005) and the ring details led to the discovery that at least 130 American Kestrels are registered as being in captivity in Britain (cf. Anonymous 2005). Furthermore, specimens collected in May 1882 in North Yorkshire, Britain, and before 1899 in Leicestershire, Britain, are not accepted onto the British list (British Ornithologists' Union 2002).

June 1976, Bearah Tor, Bodmin Moor, Cornwall, Britain. © Nigel Tucker

SPECIES ACCOUNTS

7 records

Azores (3)
1. • 17 February 1968, Achada, Terceira, male, collected (specimen location unknown) [Le Grand 1983]
2. • 1970, São Miguel, male, collected (specimen location unknown) [Le Grand 1983]
3. • 15 days in March 1980, Arrives, São Miguel, female [Le Grand 1983]

Britain (2)
1. • 25–27 May 1976, Fair Isle, Shetland, male [British Ornithologists' Union 1978; Rogers & Rarities Committee 1978; Taylor 1981]
2. • 13–28 June 1976, Bearah Tor, Bodmin Moor, Cornwall, female [British Ornithologists' Union 1978; Mellow & Maker 1981; Rogers & Rarities Committee 1978, 1979]

Denmark (1)
1. • autumn of 1901, Birkendegård, Kalundborg, Sjælland, first-year male, collected (specimen location unknown) [Donark 1953; Dybbro 1978]

Malta (1)
1. • 14 October 1967, Tal-Handaq, l/o Qormi, first-winter female, collected (private collection) [Sultana 1968]

Amur Falcon *Falco amurensis*

Breeds from south-eastern Siberia and north-eastern Mongolia east to Amurland and south to northern and eastern China and North Korea. Winters in southern Africa, from Malawi to Transvaal (del Hoyo *et al.* 1994).

A report in 1984 from both France and Britain is not genuine because the bird had escaped from a falconer (cf. CAF 2006; Grantham 2005). A falcon (probably, a first-winter male) shot in the summer of 2001 in Latvia was first thought to be this species but turned out to be either an odd Red-footed Falcon *F. vespertinus* or a hybrid (Agris Celmiņš *in litt*).

6 October 2008, Tophill Low, East Yorkshire, Britain. © Dave Bryan

1 July 2005, Jászberény, Kalmar airport, Småland, Sweden. © Lasse Olsen

6 records

Britain (1)
1 • 14 September to 15 October 2008, Tophill Low, East Yorkshire, second-winter male [Hudson & Rarities Committee; Mansell 2008; British Ornithologists' Union 2011]

Hungary (1)
1 • 10–13 July 2006, Jászberény, Borsóhalmi-legelő, first-summer female [MME NB 2010; Tamás 2008]

Italy (3)
1 • 29 April 1995, Pilone di Cannitello, Straits of Messina, Calabria, adult male [Brichetti *et al.* 1996; Corso & Dennis 1998]
2 • 04 May 1997, Straits of Messina, Sicily, female [Brichetti *et al.* 2002; Corso & Dennis 1998]
3 • 19 May 1998, Straits of Messina, Sicily, adult male [Brichetti *et al.* 2002; Corso & Dennis 1998]

Sweden (1)
1 • 30 June to 01 July 2005, Kalmar airport, Småland, adult male [Blomdahl 2006]

SPECIES ACCOUNTS

10 July 2006, Jászberény, Borsóhalmi-legelő, Hungary. © János Oláh

19 May 1998, Straits of Messina, Sicily, Italy. © Andrea Corso

RAILS Rallidae (n=3)

Striped Crake *Aenigmatolimnas marginalis*

Breeds in Africa, patchily from northern Ivory Coast east to Cameroon and south to coastal Congo. More continuously from eastern Zaire to Kenya, south and west to Zimbabwe. Also occurs at scattered locations in South Africa (del Hoyo *et al.* 1996; Taylor & van Perlo 1998).

A report in Morocco in November 1989 (De Faveri *et al.* 1998) was not accepted by Thévenot *et al.* (2003). A report of a female at Wadi Turghat, Libya, in February 1970 (Cramp & Simmons 1980) was not accepted because of insufficient documentation (Graham Bundy in litt).

4 records

Algeria (1)

1 • January 1867, Biskra, juvenile, collected (specimen location unknown) [Isenmann & Moali 2000]

Italy (1)

1 • 04 January 1997, Livorno, Toscana, adult female, taken into care, died (05 January) (INFS: 8776) [Brichetti *et al.* 1997; De Faveri *et al.* 1998]

Malta (2)

1 • 29 March 1981, Bahar ic-Caghaq, collected (NMNH: NMNH-Orn.1554) [Borg in press; Raymond Galea in litt]

2 • April 2004, Siggiewi, collected (private collection) [Raymond Galea in litt]

4 January 1997, Livorno, Toscana, Italy.
© Adriano De Faveri

SPECIES ACCOUNTS

29 March 1981, Bahar ic-Caghaq, Malta. © J. Borg

April 2004, Siggiewi, Malta. © Raymond Galea

African Crake *Crex egregia*

Breeds in sub-Saharan Africa, from Senegal east to Kenya and south to Cape Province, South Africa, except for arid areas of south-western and southern Africa (del Hoyo *et al.* 1996; Taylor & van Perlo 1998).

All records were in winter, as for Allen's Gallinule *Porphyrio alleni*, which breeds in sub-Saharan Africa and has also occurred in Europe. The first for Morocco was photographed in December 2009 (Arnoud van den Berg in litt).

5 records

Canary Islands (4)

1 • 23 November 2001, Parque García Sanabria, Santa Cruz, Tenerife, adult, taken into care and died (24 November) (MCNT) [de Juana 2006; de Juana & Comité de Rarezas de la SEO 2003; Lorenzo 2002]

2 • 15 November 2006, Radazul harbour, El Rosario, Tenerife, adult, taken into care and died (MNHT) [Díes *et al.* 2008]

3 • 16 November 2006, Almeida, Santa Cruz, Tenerife, adult, taken into care and died (MNHT) [Díes *et al.* 2008]

4 • 12 January 2007, Telde, La Garita, Gran Canaria, adult, taken into care and released (24 January) [Díes *et al.* 2009]

Cape Verde (1)

1 • 04 February 2004, Ilhéu dos Passaros, off north-eastern Boavista, adult, found dead (skeleton in BDGC) [Hazevoet 2010]

23 November 2001, Parque García Sanabria, Santa Cruz, Tenerife, Canary Islands. © Centro de Recuperación de Fauna del Cabildo Insular de Tenerife

SPECIES ACCOUNTS

12 January 2007, Telde, La Garita, Gran Canaria, Canary Islands. © Antonio Hernández García

15 November 2006, Radazul harbour, El Rosario, Tenerife, Canary Islands & 16 November 2006, Almeida, Santa Cruz, Tenerife, Canary Islands. © Juan Antonio Lorenzo

4 February 2004, Ilhéu dos Passaros, off northeast Boavista, Cape Verde. © Pedro López Suárez

73

Lesser Moorhen *Gallinula angulata*

Breeds in Africa, from Senegal and Gambia east to Ethiopia and south to northern and eastern Namibia, Botswana and eastern South Africa (del Hoyo *et al.* 1996; Taylor & van Perlo 1998).

The record on Madeira was previously published as a Black Crake *Amaurornis flavirostris* (cf. Bannerman & Bannerman 1965; Cramp & Simmons 1980; Matias 2009). Haavisto & Strand (2000) published a record of a bird photographed in Egypt in May 1997, but this has recently been reviewed by the Egyptian rarities committee and considered to be a Common Moorhen *G. chloropus* (Frédéric Jiguet / EORC in litt). A Spanish record in March 2003 was placed in category D (de Juana & Comité de Rarezas de la SEO 2005). Clarke (2006) mentioned an additional report in January 1997 from Gran Canaria, Canary Islands, but this has not (yet) been discussed by the Spanish rarities committee.

1 record

Madeira (1)
1 • 26 January 1895, Santo Amaro, Funchal, male, collected (MJBF) [Matias 2009]

CRANES Gruidae (n=1)

Sandhill Crane *Grus canadensis*

Breeds in north-eastern Siberia and North America, from Alaska and Canada south locally to northern California, Nevada, Wyoming, Colorado, South Dakota and Michigan; also from southern Mississippi, Alabama and Georgia, south through Florida to Cuba. Northern populations winter in southern North America and Mexico (Cramp & Simmons 1980; del Hoyo *et al*. 1996). It was also recorded in Greenland in July–August 1985, August 1990 and October 2007 (Boertmann 1994; Nielsen & Thorup 2001; van den Berg & Haas 2007).

The 1991 Shetland bird was relocated in the Netherlands (van den Berg *et al*. 1993). The third for Britain occurred in September 2009 (Higson 2009; Hudson & Rarities Committee 2010). Probably the same bird was recorded in France in October 2009 (Gantlett 2010).

June/July 2000, Ponta Delgada das Flores, Flores, Azores. © Hanno Schäfer

Extremely rare birds in the Western Palearctic

Sandhill Crane *Grus canadensis*

5 records

Azores (1)
1 • 26 June to 03 July 2000, Ponta Delgada das Flores, Flores, first-summer [Costa *et al.* 2003]

Britain (2)
1 • 26–27 April 1981, Fair Isle, Shetland, first-summer [British Ornithologists' Union 1984; Riddiford 1983; Rogers & Rarities Committee 1982]
2 • 17–27 September 1991, Exnaboe, Sumburgh, Shetland, first-summer [Ellis 1991; Rogers & Rarities Committee 1992] [1]

Faeroes (1)
1 • 14 October 1980, Akraberg, Suðuroy, adult female, collected (private collection) [Boertmann *et al.* 1986]

Ireland (1)
1 • c 11–14 September 1905, Galley Head, Cork, first-summer, collected (NMI: NMINH 1907.78.1) [British Ornithologists' Union 1972; Editors 1972; Hutchinson 1989; Nichols 1907; Ruttledge 1971; Whitherby 1907]

Netherlands (1)
1 • 28–30 September 1991, Paesens-Moddergat, Dongeradeel, Friesland, first-summer [van den Berg *et al.* 1993; van den Berg *et al.* 1993; van den Berg & Bosman 2001; de By *et al.* 1993] [1]

[1] Same individual

April 1981, Fair Isle, Shetland, Britain.
© J. F. Holloway

14 October 1980, Akraberg, Suðuroy, Faeroes.
© Silas Olofson

29 September 1991, Paesens-Moddergat, Dongeradeel, Friesland, Netherlands.
© Arnoud B. van den Berg

SPECIES ACCOUNTS

September 1905, Galley Head, Cork, Ireland. © Nigel Monaghan

PLOVERS
Charadriidae (n=3)

Three-banded Plover *Charadrius tricollaris*

Breeds in Africa, from Ethiopia to Tanzania and Gabon and south to South Africa, also on Madagascar. Mainly resident but some winter around Lake Chad (del Hoyo *et al*. 1996).

In April 2009, the first breeding record for the Western Palearctic occurred when two adults and two young were photographed (cf. Haas *et al*. 2010). In 2010, breeding occurred again at this site (van den Berg & Haas 2010c).

24 February 2008, Aswan, Egypt. © Knut S. Olsen

5 records (8 individuals)

Egypt (5/8)

1. • 05–26 March 1993, Gebel Asfar, Cairo [Hoath 2000]
2. • 14 December 1997, Aswan [Davidson & Kirwan 1998]
3. • 20 September 2000, Wadi El Naturn [Dave Farrow *in litt*]
4. • 13 March 2003, El Gouna golf course, Red Sea [Tomas Haraldsson *in litt*]
5. • 25 January to at least 10 February 2006, West Sehel, Aswan, *maximum of two* (on 25–30 January 2006) [Haas *et al*. 2010]
   • 25 October 2007, Tut Amon, Aswan, *two*
   • 19 November 2007, Sehel, Aswan
   • 24 February 2008, Tut Amon, Aswan, *two*
   • 02 August 2008, Tut Amon, Aswan, *two*
   • 30 April 2009, Tut Amon, Aswan, *four* (two adults and two juveniles)
   • 09–10 September 2009, Tut Amon, Aswan, *maximum of three*
   • 14 September 2009, Tut Amon, Aswan, *two*

Oriental Plover *Charadrius veredus*

Breeds in southern Siberia, northern China and Mongolia. Winters in Indonesia and northern, central and southern Australia (del Hoyo *et al.* 1996). Remarkably, there is also a record in Greenland of an adult male shot in May 1948 (Boertmann 1994).

The sole Western Palearctic record is from Finland in May.

1 record

Finland (1)
1 • 25 May 2003, Alajoki, Ilmajoki, adult male [Luoto *et al.* 2004; Rannila 2003]

25 May 2003, Alajoki, Ilmajoki, Finland. © Petri Pelttari

Black-headed Lapwing *Vanellus tectus*

Breeds in sub-Saharan Africa, from Senegal to Ethiopia and Kenya (Cramp & Simmons 1983; del Hoyo et al. 1996).

The sole Western Palearctic record concerned a male shot on the Jordan/Israeli border in 1869 by J. K. Lord. A report mentioned in Gantlett (1996) of one seen perched on the razor wire of the Jordan/Israeli border in April 1995 has never been submitted (Ian Andrews in litt).

1 record

Israel (1)

1 • sine dato, 1869, southern 'wady-el-Arabh' (Arava valley), male, collected (NHM: 1896.7.1.184) [Shirihai 1996] [1]

Jordan (1)

1 • sine dato, 1869, southern 'wady-el-Arabh' (Arava valley), male, collected (NHM: 1896.7.1.184) [Shirihai 1996] [1]

[1] Same individual

1869, southern 'wady-el-Arabh' (Arava Valley), Israel/Jordan. © Natural History Museum

SANDPIPERS
Scolopacidae (n=8)

Long-toed Stint *Calidris subminuta*

Breeds in Siberia, from Ob river to Anadyrskaya, Bering Island, northern Kurils and northern coast of Sea of Okhotsk. Winters from south-eastern China and Philippines to eastern India and south through Indonesia to Celebes; rare further south (Cramp & Simmons 1983; del Hoyo *et al*. 1996).

The first for the Netherlands occurred in October 2009 (Bot *et al*. 2010; Ovaa *et al*. 2010). A report from Greece in March 1991 (Handrinos & Akriotis 1996) was rejected after review (Hellenic Rarities Committee 2009).

7 records

Britain (2)
1 • 07–08 June 1970, Marazion, Cornwall [British Ornithologists' Union 1995; Rogers & Rarities Committee 1995; Round 1996]
2 • 28 August to 01 September 1982, Saltholme Pool, Cleveland, juvenile [British Ornithologists' Union 1986; Dunnett 1992; Rogers & Rarities Committee 1985]

June 1970, Marazion, Cornwall, Britain.
© John Johns

Extremely rare birds in the Western Palearctic

Long-toed Stint *Calidris subminuta*

Finland (1)
1 • 26–28 June 2007, Kotka, Salminlahti [Lindholm *et al.* 2008]

Ireland (1)
1 • 15–16 June 1996, Ballycotton, Cork [Milne & O'Sullivan 1997; O'Sullivan 1996]

Israel (2)
1 • 25–26 August 1991, Eilat, juvenile, trapped (25 August) [Shirihai 1996, 1999]
2 • 22–25 October 2004, Ma'oz Hayyim, Bet Shean valley [Mizrachi *et al.* 2007]

Sweden (1)
1 • 04–05 October and 15 October to 02 December 1977, Ottenby, Öland, juvenile, trapped (28 October) [Pettersson *et al.* 1978; Unger 1979]

28 June 2007, Kotka, Salminlahti, Finland. © Tero Ilomäki

22 October 2004, Ma'oz Hayyim, Bet Shean Valley, Israel. © Steve Mann

August 1982, Salthome Pool, Cleveland, Britain. © Jim Clift

Swinhoe's Snipe *Gallinago megala*

Breeds in Asia in central Siberia, northern Mongolia, Amurland and Ussuriland. Winters from southern and eastern India to eastern China and Taiwan, and south through Malay peninsula, Philippines and Indonesia to New Guinea and northern Australia (del Hoyo *et al.* 1996).

A report from the Caucasus (Cramp & Simmons 1983) was considered dubious and, therefore, not accepted. Likewise, a report from Israel in 1998 (Shirihai 1999) was not accepted (cf. Shochat *et al.* 2004). Two were recorded in Russia in June 2010 (van den Berg & Haas 2010d).

2 records

Finland (1)
1 • 13 June to 06 July 2008, Värtsilä Niirala, Tohmajärvi, male [Kivivuori *et al.* 2008; Lehikoinen *et al.* 2009]

Russia (1)
1 • 23–24 June 2002, Ynzhnaya mountains, Yengane-Pe ridge, Polar Urals, male [Morozov 2004]

June 2008, Värtsilä Niirala, Tohmajärvi, Finland. © Petri Salo

American Woodcock *Scolopax minor*

Breeds in North America, from southern Manitoba through Ontario and Quebec to New Brunswick and south-western Newfoundland, and south through Great Plains and eastern USA to eastern Texas, Gulf states and Florida. Winters in south-eastern USA, on Atlantic and Gulf coastal plains from southern Texas to southern Florida (del Hoyo *et al.* 1996).

The only Western Palearctic record consists of a bird shot in France during a hunt for Eurasian Woodcock *S. rusticola*.

1 record

France (1)
1 • 28 October 2006, Sorges-en-Périgord, Dordogne, juvenile, collected (MT) [Ferrand *et al.* 2008; Jiguet *et al.* 2009; Reeber *et al.* 2008]

28 October 2006, Sorges-en-Périgord, Dordogne, France. © Jean Brosset

SPECIES ACCOUNTS

Hudsonian Godwit *Limosa haemastica*

Breeds locally in North America, from north-western and southern Alaska to Hudson Bay. Winters on Atlantic coast of southern South America (del Hoyo et al. 1996).

A remarkable series of returning birds.

24 June 2008, Røstlandet, Nordland, Norway. © Steve Baines

25 July 2007, Cabo da Praia, Terceira, Azores. © Peter Alfrey

Extremely rare birds in the Western Palearctic

Hudsonian Godwit *Limosa haemastica*

6 records

Azores (1)
1 • 25 July 2007, Cabo da Praia, Terceira, adult [João Jara in litt]

Britain (2)
1 • 10 September to 03 October 1981, Blacktoft Sands, East Yorkshire, adult [British Ornithologists' Union 1984; Grieve 1987; Rogers & Rarities Committee 1983, 1985; Wright 1987]
• 22 November 1981 to at least 14 January 1982, Countess Wear, Devon, adult
• 26 April to 06 May 1983, Blacktoft Sands, Humberside, adult
2 • 26 September 1988, Collieston, Grampian [Rogers & Rarities Committee 1990]

Denmark (1)
1 • 06 September 1986, Kielst Bæk, Ho Bugt, c 4 km south of Oksbøl, Ribe [Olsen 1988]

Norway (1)
1 • 24 June 2008, Røstlandet, Nordland, adult male [Olsen *et al.* 2010]

Sweden (1)
1 • 14–20 June 2003, Vässby fjärd, Öland, adult male [Blomdahl & Stridh 2007; Cederroth 2004]
• 27–30 June 2006, Fyrvägen, Ottenby, Öland, adult male

6 September 1986, Kielst Bæk, Ho Bugt, Ribe, Denmark. © Lars Maltha Rasmussen

86

SPECIES ACCOUNTS

June 2006, Ottenby, Öland, Sweden.
© Christian Cederroth

November 1981, Exeter, Devon, Britain.
© Tony Croucher

87

Extremely rare birds in the Western Palearctic

Little Curlew *Numenius minutus*

Breeds in central Siberia in upper reaches of Moyero and Kochechum rivers and in north-eastern Siberia in middle reaches of Yana river and mountains to Indigirka river. Winters in New Guinea and northern Australia (Cramp & Simmons 1983; del Hoyo *et al.* 1996).

A very rare vagrant, with all records from late summer to winter.

September 1982, Sker, Mid Glamorgan, Britain. © Jim Clift

August 1985, Blakeney, Cley and Salthouse area, Norfolk, Britain. © David Cottridge

6 records

Britain (2)
1 • 30 August to 06 September 1982, Sker, Mid Glamorgan, adult [British Ornithologists' Union 1984; Moon 1983; Rogers & Rarities Committee 1983, 1985]
2 • 24 August to 03 September 1985, Blakeney, Cley and Salthouse area, Norfolk [Rogers & Rarities Committee 1986; Walker & Gregory 1987]

Finland (1)
1 • 01–02 October 1996, Haga, Saltvik, Åland, first-winter [Lindroos 1997ab; Vasamies 1997]

Kuwait (1)
1 • 12–29 December 2007, Sulaibiya Pivot Fields [Gantlett 2008; KORC 2008]

Norway (1)
1 • 14 July 1969, Varanger, Finnmark, adult [Andersson 1971]

Sweden (1)
1 • 12–13 September 2005, Ölands södra udde/Schäferiäng. Ottenby, Öland, juvenile [Blomdahl 2006]
 • 24–30 September 2005, Gräsgård - Näsbybadet, Öland, juvenile
 • 08 October to 08 November 2005, Ventlinge - Segerstads fyr, Öland, juvenile

SPECIES ACCOUNTS

October 1996, Haga, Saltvik, Åland, Finland.
© Harry Lehto

14 July 1969, Varanger, Finnmark, Norway.
© Gert Anderson

ABOVE: 15 December 2007, Sulaibiya Pivot Fields, Kuwait. © Mike Pope

BELOW: 21 October 2005, Öland, Sweden. © Stefan Pfützke

89

Eskimo Curlew *Numenius borealis*

Formerly bred in western Canada and probably in Alaska. Once wintered in South America, from Paraguay and southern Brazil to Patagonia (Cramp & Simmons 1983; del Hoyo *et al.* 1996). Now probably extinct, although still listed as critically endangered in del Hoyo *et al.* (1996); no recent records have been published. It has also been recorded five times in Greenland, with the most recent record in 1882 (Boertmann 1994).

The three specimens collected in Aberdeenshire may suggest a regular but scarce presence. The BOU records committee has recently decided to reject the record of two birds in Suffolk in November 1852 (cf. British Ornithologists' Union 2010; Melling 2010).

5 records

Britain (4)

1. • 06 September 1855, Cairn Monearn, Stonehaven, Kincardineshire, Aberdeenshire, collected (skin lost) [British Ornithologists' Union 2007a, 2010; Evans 1994; Melling 2010]
2. • 29 September 1878, Slains, Aberdeen, Aberdeenshire, male, collected (skin lost) [British Ornithologists' Union 2007a; Evans 1994; Harting 1879; Melling 2010]
3. • 21 September 1880, Forest of Birse, Kincardineshire, Aberdeenshire, adult male, collected (skin lost) [British Ornithologists' Union 2007a; Evans 1994; Harvie Brown 1880; Melling 2010; Sim 1880]
4. • 10 September 1887, Tresco, Isles of Scilly, adult, collected (ISM: RN 00956) [British Ornithologists' Union 2007a; Cornish 1887; Evans 1994; Melling 2010]

Ireland (1)

1. • 21 October 1870, obtained at Dublin market, Sligo, collected (NMI: NMINH 1884.293) [Blake-Knox 1870; Ussher & Warren 1900]

21 October 1870, Sligo, Ireland.
© Nigel Monaghan

10 September 1887, Tresco, Isles of Scilly, Britain.
© Amanda Martin / Isles of Scilly Museum

Grey-tailed Tattler *Tringa brevipes*

Breeds in northern-central and north-eastern Siberia in Putorana mountains and from Verkhoyansk mountains east to Anadyrland. Winters from Taiwan, Malay peninsula and Philippines through Indonesia, New Guinea and Solomon Islands to Australia (del Hoyo *et al.* 1996).

An extremely rare vagrant from the Eastern Palearctic.

3 records

Britain (2)
1. • 13 October to 17 November 1981, Dyfi estuary, Dyfed/Gwynedd [British Ornithologists' Union 1988; Rogers & Rarities Committee 1987; Thorpe 1995]
2. • 27 November to 27 December 1994, Burghead, Grampian, juvenile [Rogers & Rarities Committee 1995; Stenning & Hirst 1994]

Sweden (1)
1. • 12 July 2003, Sebybadet, Gammalsbyören, Öland [Cederroth 2004]

12 July 2003, Sebybadet, Gammalsbyören, Öland, Sweden. © Jens Berkler

December 1994, Burghead, Grampian, Britain. © Steve Young

Extremely rare birds in the Western Palearctic

Willet *Tringa semipalmata*

Breeds in North America, in central Alberta and south-western Manitoba south to northern Colorado and western Nebraska, and along Atlantic and Gulf coasts of North America. Winters along Pacific, Gulf and Atlantic coasts of North and South America (del Hoyo *et al.* 1996).

The second French bird was assigned to subspecies Western Willet *T. s. inornatus* (Jiguet *et al.* 2009). The first for Portugal was photographed in April 2009 (van den Berg & Haas 2009; Gantlett 2010). The third for the Azores was reported in November 2006, while the fourth turned up in September 2009 (Gantlett 2010; João Jara in litt). Le Grand (1993) mentioned two additional reports from the Azores (July 1984 and June 1985), but these have not been submitted to the Portuguese rarities committee (João Jara in litt).

7 records

Azores (2)
1 • 12 March 1979, Populo, São Miguel, found long dead (specimen location unknown) [Le Grand 1983]
2 • 22 September 2006, Fajã dos Cubres, São Jorge [João Jara in litt]

Finland (1)
1 • 21 September 1983, Kuivanuoro, Kemi, first-winter [Mikkola 1984]

France (2)
1 • sine dato, 1867, Abbeville, Somme, collected (specimen location unknown) [Dubois *et al.* 2008]
2 • 12–13 September 1998, lagune de La Belle Henriette, La Faute-sur-Mer, Vendée, first-winter [Dubois *et al.* 2008; Frémont & CHN 1999; Reeber *et al.* 2008; Siblet & Spanneut 1998]

Italy (1)
1 • 02 February 2008, Riserva Naturale Regionale di Punta Aderci, Vasto, Abruzzo, first-winter [Antonucci & Corso 2008; Janni & Fracasso 2009]

Norway (1)
1 • 14 October 1992 to 23 March 1993, Mølen and Omlidstranda, Larvik, Vestfold, first-winter [Gustad 1994, 1995a; Sondbø 1992]

October 1992, Larvik, Vestfold, Norway.
© Vidar Heibo

SPECIES ACCOUNTS

September 1998, lagune de La Belle Henriette / La Faute-sur-Mer, Vendée, France.
© Laurent Spanneut

2 February 2008, Riserva Naturale di Punta Aderci di Vasto, Abbruzo, Italy. © Antonio Antonucci

93

SKUAS
Stercorariidae (n=1)

South Polar Skua *Stercorarius maccormicki*

Breeds in Antarctica. Winters in Pacific Ocean; scarce in Atlantic (Olsen & Larsson 1997). It was also recorded in Greenland in July 1902 and July 1975 (Boertmann 1994).

A report in November 2000 from the Azores was not accepted by the Portuguese rarities committee (cf. Jara *et al*. 2007). Furthermore, two reports from the Azores in September 1996 and September 1998 (Olsen & Larsson 1997; Clarke 1999) have not (yet) been submitted to the Portuguese rarities committee (João Jara in litt). A report in Egypt of an intermediate morph between Hurghada and Giftun Saghir island on 10 June 1991 was mentioned in Baha El Din (1992) but was published without details. The Spanish rarities committee has accepted a skua off La Palma, Canary Islands, in October 2005 as a 'southern skua' *S. maccormicki/antarcticus* (Díes *et al*. 2007; Winkel 2009). In Britain, skuas in the Isles of Scilly in October 2001 and in Glamorgan in February 2002 were considered to be Brown Skuas *S. antarcticus* (Moon & Carrington 2002; Scott 2002), which was confirmed by Votier *et al*. (2004) based on DNA samples. In 2007, this was retracted by Votier *et al*. (2007) as supplementary DNA tests had failed to confirm the species' identity; although they were most likely to be South Polar Skuas, this cannot be stated with certainty (Hudson & Rarities Committee 2010; Newell 2008).

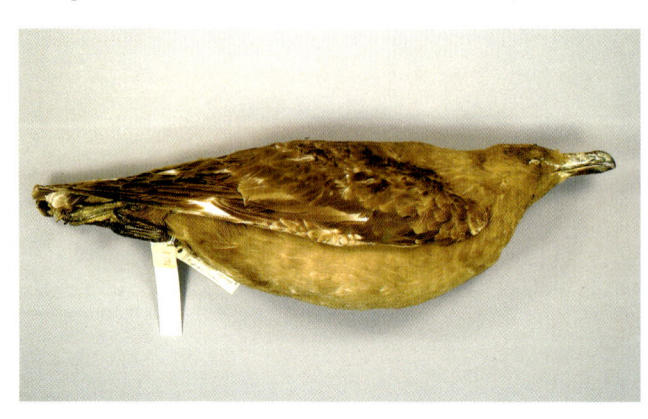

24 September 1889, no location, Faeroes.
© Margaret Hart / American Museum of Natural History

3 records

Faeroes (1)
1 • 24 September 1889, sine loco, juvenile female, collected (AMNH: AMNH 744420) [Boertmann *et al.* 1986; Bourne 1989]

Israel (2)
1 • 03–06 June 1983, Eilat, pale morph [Shirihai 1996, 1999] [1]
2 • 28 June 1992, Eilat [Shirihai 1996, 1999]

Jordan (1)
1 • 03–06 June 1983, Aqaba, pale morph [Andrews *et al.* 1999] [1]

[1] Same individual

GULLS Laridae (n=5)

Brown-headed Gull *Croicocephalus brunnicephalus*

Breeds in mountains of southern-central Asia, from Turkestan east to south-eastern Gansu and south to Pamirs, Ladakh and Tibet. Winters on coast of India, northern Sri Lanka and south-eastern Asia, west to Arabian peninsula (del Hoyo *et al.* 1996).

The identification of the record from Israel was questioned by Hoogendoorn (1991) but is still accepted by the Israeli rarities committee. In Shirihai *et al.*'s (1987) description, based on a translation of the original Hebrew description (cf. Shirihai 1999), some important features are missing.

1 record

Israel (1)

1 • 12 May 1985, Eilat, first-summer [Shirihai 1996, 1999; Shirihai *et al.* 1987]

Relict Gull *Larus relictus*

Breeds very locally in central Asia. Winters in south-eastern Asia (del Hoyo *et al.* 1996).

Two ringing recoveries from Bulgaria and Turkey in March 1978 and March 1990, respectively, are considered doubtful (cf. Gavrilov & Gavrilov 2000); therefore, the Kazakh/Russian record is the sole acceptable Western Palearctic record.

1 record (14 individuals)

Kazakhstan (1)

1 • 07 May 2000, Malyy Uzen river, West Kazakhstan, *14* (adults) [Piskunov & Antonchikov 2007; Wassink & Oreel 2007] [1]

Russia (1)

1 • 07 May 2000, Malyy Uzen river, Dryamsky, Saratov oblast, *14* (adults) [Piskunov & Antonchikov 2007] [1]

[1] Same individuals

Glaucous-winged Gull *Larus glaucescens*

Breeds from Commander Islands east through Aleutians, Pribilofs and southern Bering Sea to Alaska, and southeastern to northern Oregon. Winters from Bering Sea to northern Japan and Baja California (del Hoyo *et al.* 1996).

A record in February 1992 on El Hierro, Canary Islands, was placed in category D (cf. de Juana 2006; de Juana & Comité de Rarezas de la SEO 1998). The records from Morocco and the Canary Islands might refer to the same bird. The first record from Denmark was in November 2009; the bird reappeared briefly in December 2009 and February 2010 (van den Berg & Haas 2010ab; Gantlett 2010).

3 records

Britain (2)
1. • 15–16 December 2006, Hempsted, Gloucestershire, third-winter, trapped (15 December) [Allan 2007; British Ornithologists' Union 2010; Hudson & Rarities Committee 2009; Sanders 2007, 2010]
   • 02–05 March 2007, Ferryside, Carmarthenshire, third-winter
   • 16–17 March 2007, Hempsted, Gloucestershire, third-winter
   • 18 April 2007, Beddington sewage farm, Greater London/Surrey, third-winter

December 2006, Hempsted, Gloucestershire, Britain. © John Sanders

Extremely rare birds in the Western Palearctic

Glaucous-winged Gull *Larus glaucescens*

2 • 31 December 2008 to 10 January 2009, Saltholme, Cowpen and nearby, Cleveland, adult [Collett 2009; Hudson & Rarities Committee 2009, 2010]

Morocco (1)

1 • 31 January 1995, Ksob estuary, Essaouira, Chiadma, adult winter [Bakker *et al.* 2001; Bergier *et al.* 1997, 2000; Thévenot *et al.* 2003]

31 January 1995, Ksob estuary, Essaouira, Chiadma, Morocco. © Theo Bakker

January 2009, Saltholme, Cleveland, Britain. © Steven Gantlett

Cape Gull *Larus dominicanus*

Mainly resident in southern Africa and southern Madagascar; southern Australia and New Zealand; as well, northern Peru and southern Brazil south to Tierra del Fuego and Falkland Islands (del Hoyo *et al.* 1996).

All birds recorded in the Western Palearctic belong to the African subspecies *vetula*. A long-staying bird at Banc d'Arguin, Mauritania, in 1997–2007 constituted the first record for the Western Palearctic. The record of an adult on Tenerife, Canary Islands, in April 2001 might relate to the same bird. In May 2009, the first Western Palearctic breeding record was established in Morocco when a nest with three eggs was found (cf. Bergier *et al.* 2009). Up to eight birds were still present at the breeding site in December 2009 (Gantlett 2010). Van den Berg & Haas (2010c) mention breeding in Morocco again in April 2010 and the occurrence of another bird/pair further north in April 2010.

5 records (14 individuals)

Canary Islands (1)
1 • 15 April 2001, Arona, Punta de la Rasca, Tenerife, adult [de Juana 2006; de Juana & Comité de Rarezas de la SEO 2006]

France (1)
1 • January 1995, Jardin des Plantes, Paris, adult [Frémont & CHN 2003; Jiguet & Defos du Rau 2004; Jiguet *et al.* 2009]

Mauritania (1)
1 • 23 April 1997 to at least 21 December 2007, Zira islet, Banc d'Arguin, adult [Gantlett 2008; Isenmann 2007; Pineau *et al.* 2001]

Morocco (2/11)
1 • 28 August 2006, Agadir, Souss, adult [Bergier *et al.* 2008]
2 • 19–24 February 2008, Khnifess, Tarfaya, *maximum of four* (adults) [Bergier *et al.* 2009, 2010]
• 17–24 February 2009, Khnifess, Tarfaya, *eight* (one first-winter and seven adults)
• 15 April to 20 May 2009, Khnifess, Tarfaya, *maximum of 10* (adults), two pairs breeding, one nest containing three eggs

8 December 2007, Zira islet, Banc d'Arguin, Mauritania. © Jochen Dierschke

Extremely rare birds in the Western Palearctic

Cape Gull *Larus dominicanus*

January 1995, Jardin des Plantes, Paris, France. © Frédéric Jiguet

28 August 2006, Agadir, Souss, Morocco. © Frédéric Jiguet

SPECIES ACCOUNTS

Slaty-backed Gull *Larus schistisagus*

Breeds in Asia, in north-eastern Siberia, from Cape Navarin and Kamchatka south to Hokkaido, north-eastern Honshu and Vladivostok. Winters south to Taiwan (del Hoyo *et al.* 1996).

The Lithuanian bird was also found in Latvia in April 2009 (van den Berg & Haas 2009; Gantlett 2010; Gibbins & Hackett 2009).

1 record

Lithuania (1)
1 • 17–18 November 2008, Klaipeda rubbish dump, near-adult [Gantlett 2009]

November 2008, Klaipeda rubbish dump, Lithuania. © Vytautas Jusys

101

TERNS Sternidae (n=3)

Aleutian Tern *Onychoprion aleuticus*

Breeds from Sakhalin through Sea of Okhotsk to Kamchatka; Alaska, on Bering and Pacific coasts, from Chukchi Sea and Kotzebue Sound south to Aleutian Island and east to Copper River delta and Dry Bay. Winter area largely still unknown, although some winter in Philippines and Hongkong (del Hoyo *et al.* 1996).

A remarkable record because at the time it had never been seen outside its known breeding area (cf. Dixey *et al.* 1981).

1 record

Britain (1)

1 • 28–29 May 1979, Farne Islands, Northumberland [British Ornithologists' Union 1984; Dixey *et al.* 1981; Rogers & Rarities Committee 1980]

May 1979, Farne Islands, Northumberland, Britain. © A. R. Taylor

Least Tern *Sternula antillarum*

Breeds in central California to Baja California and western Mexico; inland rivers in central North America, from northern Great Plains to northern Louisiana and Texas; Bermuda, and Maine south to Texas and Honduras and through Caribbean to northern Venezuela. Winters in Central America to northern Brazil (del Hoyo *et al.* 1996).

The sole record is of a bird that spent nearly 10 summers in Britain. It took the BOU records committee a long time to accept this record because the Little Tern subspecies *S. albifrons guineae* breeding in Africa could not be ruled out, mainly because little information on its calls was available (cf. British Ornithologists' Union 2006a; Yates 2010).

1 record

Britain (1)

1. • c 08 June to 05 July 1983, Rye Harbour, East Sussex, male [British Ornithologists' Union 2006ab; Evans 1994; Rogers & Rarities Committee 2005; Yates 2010; Yates & Taffs 1990]
mid-May to late June 1984, Rye Harbour, East Sussex, male
• 30 May 1985, Rye Harbour, East Sussex, male
• 22 May 1986, Rye Harbour, East Sussex, male
mid-May to late June 1987, Rye Harbour, East Sussex, male
• 23 May to 08 June 1988, Rye Harbour, East Sussex, male
• 28 May to 24 June 1989, Rye Harbour, East Sussex, male
• 21 May to 12 July 1990, Rye Harbour, East Sussex, male
• 31 May to 01 July 1990, Colne Point, Essex, male
• 25 May to 13 July 1991, Rye Harbour, East Sussex, male
• 24 May to 12 July 1992, Rye Harbour, East Sussex, male

Brown Noddy *Anous stolidus*

Breeds on Caribbean and South Atlantic islands and on islands in Indian and Pacific Oceans (del Hoyo *et al.* 1996).

A report in Norway in August 1974 was initially accepted as Brown Noddy (cf. Michaelsen 1985), but was subsequently changed to noddy *Anous* sp. after a revision in 2002 (Mjøs 2002). Hazevoet (1985) published a report of a Black Noddy *A. minutus* in Mauritania in December 1984, but its identification has been questioned by Beaman (1990); this same report was published by Isenmann (2007) as a possible Black Noddy. In August 2008, a Brown Noddy was photographed 200 nautical miles off Flores, Azores, outside Western Palearctic waters.

1 record

Germany (1)
1 • early October 1912, Simonsberg, Schleswig-Holstein, collected (NMF) [von Niethammer *et al.* 1964; Paulsen 1925]

Early October 1912, Simonsberg, Schleswig-Holstein, Germany. © Klaus Günther

SPECIES ACCOUNTS

AUKS Alcidae (n=5)

Long-billed Murrelet *Brachyrhampus perdix*

Breeds in Kamchatka and Sea of Okhotsk south to northern Japan. Winters offshore and in open seas within breeding range (del Hoyo *et al.* 1996).

The two records in 2006 are of different birds.

Between 15 and 18 December 1997, Zürichsee, Zollikon, Zürich, Switzerland. © Peter Knaus

3 records

Britain (1)
1 • 07–14 November 2006, Dawlish, Devon, juvenile [British Ornithologists' Union 2007b; Fraser & Rarities Committee 2007; Hopkins *et al.* 2006; Rylands 2008]

Romania (1)
1 • 21–23 December 2006, Porumbacu de Jos, Transylvania, juvenile [Gantlett 2007]

Switzerland (1)
1 • between 15 and 18 December 1997, Zürichsee, Zollikon, Zürich, first-winter, found dead in a fishnet (NHMB: 99-004) [Knaus & Balzari 1999; Knaus *et al.* 2000; Maumary & Knaus 2000]

105

Extremely rare birds in the Western Palearctic

Long-billed Murrelet *Brachyrhampus perdix*

November 2006, Dawlish, Devon, Britain. © Steve Young

23 December 2006, Porumbacu de Jos, Transylvania, Romania. © János Oláh

Ancient Murrelet *Synthliboramphus antiquus*

Breeds in Yellow Sea and Sea of Japan north through Sakhalin, Kuril Island and northern Sea of Okhotsk to Kamchatka, east through Aleutians to southern Alaska and to Queen Charlotte Island. Winters offshore and in open seas within breeding range to northern California (del Hoyo *et al.* 1996).

Remarkably, the British bird returned in three consecutive springs.

1 record

Britain (1)

1 • 27 May to 26 June 1990, Lundy, Devon, adult [British Ornithologists' Union 1992a; Campey & Mortimer 1990; Rogers & Rarities Committee 1992, 1993; Waldon 1994]
• 14 April to at least 20 June 1991, Lundy, Devon, adult
• 30 March to 29 April 1992, Lundy, Devon, adult

17 June 1990, Lundy, Devon, Britain.
© Paul Hopkins

Crested Auklet *Aethia cristatella*

Breeds in Siberia (Chukotskiy peninsula), islands in the Bering Sea, central Kuril Islands and in Alaska on Aleutians, Shumagin Islands and Kodiak Islands. Winters offshore and in open seas within breeding range, south to northern Japan (Cramp 1985; del Hoyo *et al.* 1996). It was also recorded once in Greenland between 1968 and 1972 (Boertmann 1994).

Yet another remarkable record of a Pacific auk.

1 record

Iceland (1)

1 • between 12 and 20 August 1912, c 45 nautical miles east-northeast off Langanestá, 66° 48′ N 12° 55′ W, at sea, adult, collected (ZM: ZM 2.3.1934.CN.1) [Hørring 1933; Pétursson 1987; Pétursson & Þráinsson 1999]

Between 12 and 20 August 1912, off Langanestá, at sea, Iceland. © Joakim Engel

Parakeet Auklet *Aethia psittacula*

Breeds from Chukotskiy peninsula south to Commander Islands, Kamchatka, Sea of Okhotsk and Kuril Islands, and western Alaska south to Aleutians and east to northern Gulf of Alaska. Winters offshore and in open seas from Bering Sea south to northern Japan (del Hoyo *et al.* 1996).

A remarkable record of a bird from an inland lake.

1 record

Sweden (1)
1 • mid-December 1860, Vättern, Jönköping, Småland, found dead (SMNH: A537036) [Breife *et al.* 2003]

Mid-December 1860, Vättern, Jönköping, Småland, Sweden. © Göran Frisk

Tufted Puffin *Fratercula cirrhata*

Breeds from northern Japan through Kuril Islands, Sakhalin and Sea of Okhotsk to northern Chukotskiy peninsula, and western Alaska south to Aleutians, east through Gulf of Alaska and south to British Columbia and central California. Winters at sea and along coasts from Kamchatka and southern Alaska through breeding range (del Hoyo *et al.* 1996). Possibly, the same bird as the first for Britain in 2009 was photographed in Greenland in August 2009 (van den Berg & Haas 2010a).

The first for Britain was photographed in September 2009 (British Ornithologists' Union 2011, Hudson & Rarities Committee 2010; Wright 2009).

1 record

Sweden (1)
1 • 01 and 08 June 1994, Lagans mynning, Laholmsbukten, Halland, adult [Cederroth 1995]

109

Extremely rare birds in the Western Palearctic

DOVES
Columbidae (n=2)

Yellow-eyed Pigeon *Columba eversmanni*

Breeds in western Asia, from Aral Sea east to Tien Shan and south to north-eastern Iran and Afghanistan. Winters south to south-eastern Iran, Pakistan and north-western India. Most populations are migratory, moving south to winter (del Hoyo *et al.* 1997).

Arkhipov *et al.* (2010) mention three additional reports in Russia but these have not been accepted.

1 record

Russia (1)
1 • 12 May 1881, Orenburg, adult, collected (ZMSP: 31440) [Arkhipov *et al.* 2010]

12 May 1881, Orenburg, Russia.
© Vladimir Loskot

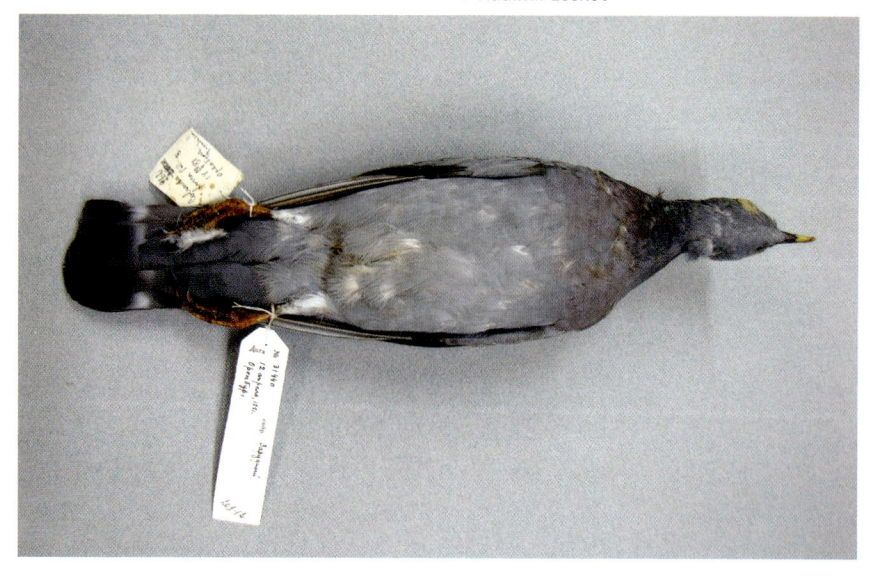

Mourning Dove *Zenaida macroura*

Breeds in North America, from British Columbia east to Nova Scotia and south to California and Florida; also Costa Rica and western Panama. Resident in central and southern North America, but northern populations move south to winter in Central America (del Hoyo *et al.* 1997). Birds were also recorded in Greenland in July 1901 and October 1946 (Boertmann 1994).

Remarkably, photographic evidence shows that the bird seen in Ireland in November 2007 and in Germany and Denmark in May 2008 was the same. A record from Sweden in June 2001 was placed in category D (Cederroth 2002). There is also a report of a bird found at Heathrow Airport, London, Britain, in February 1998, having arrived on an airplane from Chicago, USA (Anonymous 1998b). The second for the Azores was reported in October 2008 (João Jara *in litt*). The second for Ireland occurred in October 2009 (Millington 2009).

2 November 2005, Vila Novo do Corvo, Corvo, Azores. © Peter Alfrey

31 October 1989, Calf of Man, Isle of Man, Britain. © Ian Fisher

Extremely rare birds in the Western Palearctic

Mourning Dove *Zenaida macroura*

7 records

Azores (2)
1 • 02 November 2005, Vila Novo do Corvo, Corvo [Jara *et al.* 2007]

Britain (3)
1 • 31 October to 01 November 1989, Calf of Man, Isle of Man, taken into care and died (01 November) (MM) [British Ornithologists' Union 1993a; Rogers & Rarities Committee 1993, 1996; Sapsford 1990, 1996]
2 • 13–15 November 1999, Carinish, North Uist, Outer Hebrides, first-winter [British Ornithologists' Union 2003; Rabbitts 1999, 2008; Rogers & Rarities Committee 2000]
3 • 29 October to 07 November 2007, Carnach, North Uist, Outer Hebrides, first-winter [Hudson & Rarities Committee 2008; Rabbitts 2007]

Denmark (1)
1 • 19–21 May 2008, Skagen, Nordjylland [Kristensen *et al.* 2009] [1]

Germany (1)
1 • 04 May 2008, Greifswalder Oie, Mecklenburg-Vorpommern, adult [Deutsche Seltenheitenkommission 2009] [1]

Iceland (1)
1 • 19 October 1995, Heimaey, Vestmannaeyjar, immature female, collected (private collection) [Pétursson 1996; Þráinsson & Pétursson 1997]

Ireland (1)
1 • 02–15 November 2007, Inishbofin Island, Galway, first-winter [Milne & McAdams 2009] [1]

[1] Same individual

19 May 2008, Skagen, Nordjylland, Denmark. © Ole Krogh

SPECIES ACCOUNTS

November 1999, Carinish, North Uist, Outer Hebrides, Britain. © Nic Hallam

19 October 1995, Heimaey, Vestmannaeyjar, Iceland. © Ingvar A. Sigurðsson

6 November 2007, Inishbofin Island, Galway, Ireland. © Anthony McGeehan

4 May 2008, Greifswalder Oie, Mecklenburg-Vorpommern, Germany. © Mathias Mähler

November 2007, Carnach, North Uist, Outer Hebrides, Britain. © Gary Jenkins

CUCKOOS
Cuculidae (n=2)

Jacobin Cuckoo *Clamator jacobinus*

Breeds in sub-Saharan Africa, from Senegal east to Red Sea and south to South Africa. Also in south-eastern Iran, Pakistan and India. Some populations show dispersive movements (Cramp 1985; del Hoyo *et al.* 1997).

A record from Finland in September 1976 has been placed in category E (Antero Lindholm in litt).

1 record

Chad (1)
1 • 09 September 1955, Aouzou, Tibesti, collected (specimen location unknown) [Malbrant 1957]

Dideric Cuckoo *Chrysococcyx caprius*

Breeds in sub-Saharan Africa, from Senegal east to Ethiopia and south to South Africa. Northern populations migrate (Cramp 1985; del Hoyo *et al.* 1997). It also breeds in south-western Arabia, where it is a fairly common breeding visitor (Eriksen *et al.* 2003).

The Israeli bird was seen by many birders.

2 records

Cyprus (1)
1 • 27 June 1982, Akrotiri, adult female [Cyprus Ornithological Society (1957) 1983; Lobb 1983]

Israel (1)
1 • 13–26 March 1994, Eilat, male [Shirihai 1996]

March 1994, Eilat, Israel. © Stefan Pfützke

SWIFTS Apodidae (n=1)

Fork-tailed Swift *Apus pacificus*

Breeds from Siberia east to Kamchatka and Japan and south through China to Thailand, Burma, outer Himalayas and Khasai hills. Northern populations winter in Malaysia south to Australia (Chantler & Driessens 2000; Cramp 1985; del Hoyo *et al.* 1999).

All records are from late May–August and coincide with occurrences of White-throated Needletail *Hirundapus caudacutus* in the Western Palearctic, which has similar breeding and wintering ranges (Sanders *et al.* 1998).

8 records

Britain (5)
1 • 19 June 1981, North Sea Shell B. T. gas platform, Leman Bank, 53° 06 N 02° 12 E, c 45 km north-east of Happisburgh, Norfolk, trapped, released (Beccles, Norfolk; 19 June) [British Ornithologists' Union 1984, 1988; Parker 1990; Rogers & Rarities Committee 1983]
• 20 June 1981, Shadingfield area, Norfolk

30 May 1993, Cley, Norfolk, Britain. © Steve Young

Extremely rare birds in the Western Palearctic

Fork-tailed Swift *Apus pacificus*

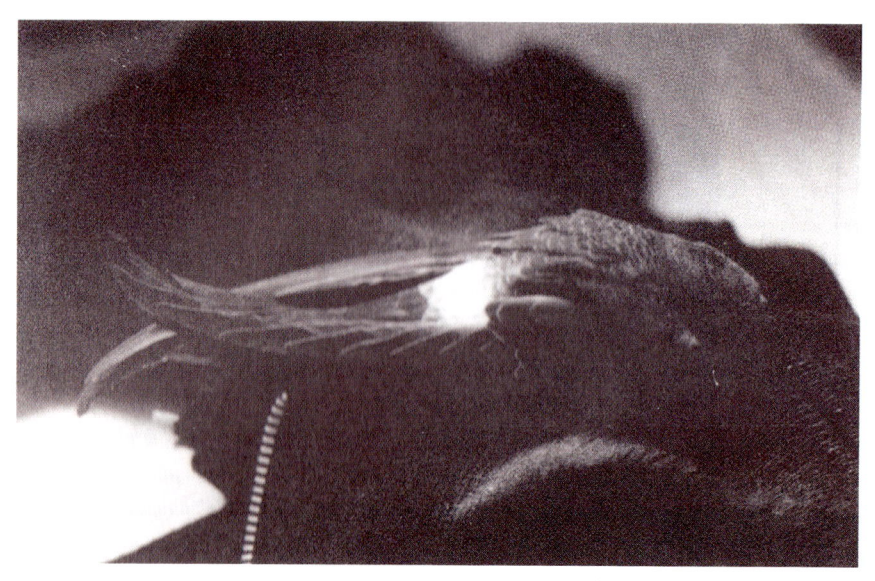

19 June 1981, off Happisburg, Norfolk, Britain. © Gary Davies

2 • 30 May 1993, Cley, Norfolk [Gantlett 1993; Rogers & Rarities Committee 1994]
3 • 16 July 1995, Daventry Reservoir, Northamptonshire [Rogers & Rarities Committee 1996]
4 • 01 July 2005, Spurn, East Yorkshire [Fraser *et al.* 2007a]
5 • 22 June 2008, Beacon Ponds, Kilnsea, East Yorkshire [Hudson & Rarities Committee 2009]
 • 26 June 2008, Spurn and Kilnsea, East Yorkshire

Sweden (3)
1 • 06 July 1999, Getteröns naturreservat, Halland [Cederroth 2000]
2 • 30 July 2005, Hoburgen, Gotland [Blomdahl 2006]
3 • 19 August 2007, Krákenabben & Hällevik, Blekinge [Hellström & Strid 2008]

African Palm Swift *Cypsiurus parvus*

Breeds in sub-Saharan Africa, from Senegal east to north-western Ethiopia and southern Somalia, south to Angola, central Namibia, northern Botswana and into South Africa. It also breeds on Madagascar and Comoros. In Arabia, a population occurs along coast and Red Sea mountains in Asir region of Saudi Arabia and into Yemen (Cramp 1985; Chantler & Driessens 2000; del Hoyo *et al.* 1999).

Haas *et al.* (2010) listed all Western Palearctic records (including a few from Egypt) and concluded that only the recent record from Mauritania was acceptable.

1 record (2 individuals)

Mauritania (1/2)
1 •07 July 1999, Gueltet El Begra, south of Char palm grove, *two* [Haas *et al.* 2010]

ROLLERS
Coraciidae (n=1)

Broad-billed Roller *Eurystomus glaucurus*

Breeds in sub-Saharan Africa, from Senegal east to Eritrea and south (except forest zones) to northern Transvaal, northern Zululand, northern Zimbabwe and Angola; also on Madagascar (Cramp 1985; Fry *et al.* 1992; del Hoyo *et al.* 2001).

Both specimens belong to the northern subspecies *afer* (Hazevoet 1995).

2 records

Cape Verde (2)

1 • 22 November 1897, Maio, juvenile male, collected (NHM: 1911.12.23.566) [Hazevoet 1995]
2 • 28 April 1924, Praia, Santiago, female, collected (CMNH: 2455) [Hazevoet 1995]

22 November 1897, Maio, Cape Verde. © Natural History Museum

28 April 1924, Praia, Santiago, Cape Verde. © Andrew W. Jones / The Cleveland Museum of Natural History

WOODPECKERS
Picidae (n=2)

Northern Flicker *Colaptes auratus*

Breeds in North America, from treeline in Alaska and Canada south to Gulf of Mexico and Greater Antilles (Cramp 1985; del Hoyo *et al.* 2002).

The Danish record is of a bird photographed in a garden near a harbour and, obviously, there is a good possibility of a ship-assisted crossing of the Atlantic. A decomposed corpse was found on board a ship that arrived in Caithness, Britain, in July 1981. Supposedly, the bird died outside of European waters and was not accepted to the British list (cf. Rogers & Rarities Committee 1982). One of the 10 or more that landed on board a ship during a crossing from New York, USA, to Southampton, Hampshire, Britain, survived and was seen flying ashore just inside Cobh Harbour, Cork, Ireland, in October 1962 (Durand 1963). Because of ship assistance and being fed during the crossing, it was not admitted to the Irish list (cf. British Ornithologists' Union 1972, 2010; Evans 1994).

1 record

Denmark (1)
1 • 18 May 1972, Ålborg, Nordjylland, female [Hansen *et al.* 1974]

18 May 1972, Ålborg, Nordjylland, Denmark. © Hans Krabsen

Yellow-bellied Sapsucker *Sphyrapicus varius*

Breeds in North America, from eastern Canada west to Alberta, north-eastern British Columbia, southern Yukon, South Dakota, Illinois, Ohio, Pennsylvania and New York, with some southward extension in Allegheny mountains. Winters in central and southern North America, Central America and Greater Antilles (Cramp 1985; del Hoyo *et al.* 2002).

It was also recorded from Greenland in July 1845, c 1857–58, the spring of 1926 and May 1934 (Boertmann 1994).

29 October 2008, Corvo, Azores. © Stefan Pfützke

September 1975, Tresco, Isles of Scilly, Britain. © David Hunt

October 1988, Cape Clear Island, Cork, Ireland. © Anthony McGeehan

SPECIES ACCOUNTS

5 records

Azores (1)
1 • 11 October to 03 November 2008, Corvo, first-winter male [João Jara in litt]

Britain (1)
1 • 26 September to 06 October 1975, Tresco, Isles of Scilly, immature male [British Ornithologists' Union 1978; Dymond & Rarities Committee 1976; Hunt 1979]

Iceland (2)
1 • 05 June 1961, Fagurhólsmýri í Öræfum, Austur-Skaftafellssýsla, adult female, found dead (IMNH: RM4059) [Ólafsson 1993a; Pétursson & Þráinsson 1999]
2 • 07–13 October 2007, Selfoss, Árnessýsla, first-winter male [Yann Kolbeinsson in litt]

Ireland (1)
1 • 16–19 October 1988, Cape Clear Island, Cork, immature female, trapped (16 October) [O'Sullivan & Smiddy 1989; Watmough 1988]

8 October 2007, Selfoss, Árnessýsla, Iceland. © Örn Óskarsson

5 June 1961, Fagurhólsmýri í Öræfum, Austur-Skaftafellssýsla, Iceland. © Yann Kolbeinsson

121

TYRANT-FLYCATCHERS
Tyrannidae (n=5)

Eastern Phoebe *Sayornis phoebe*

Breeds in North America, from north-western Canada east to Maritime Provinces, and from Great Plains south to central Texas and east to Atlantic coast south to Georgia. Winters in south-eastern and southern USA south to southern Mexico (del Hoyo *et al.* 2004).

Interestingly, another bird was reported in Devon, Britain, in April 1987, just two days before the Lundy record (cf. Croft & Davies 1987; McShane 1996).

1 record

Britain (1)
1 • 24–25 April 1987, Lundy, Devon [British Ornithologists' Union 1993a; McShane 1996; Rogers & Rarities Committee 1993]

SPECIES ACCOUNTS

Acadian Flycatcher *Empidonax virescens*

Breeds in North America, from southeastern North Dakota, southern Michigan, extreme south of Ontario, southern New York and south-western Connecticut south to central Texas, Gulf coast and central Florida. Winters from Nicaragua to Ecuador and Venezuela (Cramp 1988; del Hoyo *et al.* 2004).

Three *Empidonax* species have been recorded in the Western Palearctic, namely Acadian, Alder and Least Flycatchers in Iceland and Alder Flycatcher in Britain.

1 record

Iceland (1)
1 • 04 November 1967, Selfoss, Árnessýsla, found dead (IMNH: RM4063) [Kolbeinsson *et al.* 2006; Ólafsson 1993a; Pétursson & Þráinsson 1999]

4 November 1967, Selfoss, Árnessýsla, Iceland.
© Yann Kolbeinsson

123

Extremely rare birds in the Western Palearctic

Alder Flycatcher *Empidonax alnorum*

Breeds in North America, from western Alaska south to British Columbia and east to Canadian Maritime Provinces, and from Great Lakes east to New England and south to northern Indiana, Ohio and North Carolina. Winters mainly in western South America, from central Colombia and eastern Ecuador south to eastern Peru and Bolivia (del Hoyo *et al*. 2004).

The first bird for Britain was trapped in October 2008 (Wilson 2008). Identification is very difficult because of its close resemblance to Willow Flycatcher *E. traillii*, but both this and the Icelandic bird were trapped and their biometrics supported identification as Alder Flycatcher.

1 record

Iceland (1)
1   • 11 October 2003, Seljaland undir Eyjafjöllum, Rangárvallasýsla, trapped [Kolbeinsson *et al*. 2006, 2006]

11 October 2003, Seljaland undir Eyjafjöllum, Rangárvallasýsla, Iceland. © Daniel Bergmann

Least Flycatcher *Empidonax minimus*

Breeds in North America, from central Canada south in USA to north-eastern Colorado, northern Great Plains, New England and north-western Georgia. Winters in Central America (del Hoyo *et al.* 2004).

This record came less than a week before the capture of the Alder Flycatcher in Iceland.

1 record

Iceland (1)
1 • 06 October 2003, Stokkseyri, Árnessýsla, flew into a house, released (07 October) [Kolbeinsson *et al.* 2006, 2006]

6 October 2003, Stokkseyri, Árnessýsla, Iceland. © Jóhann Óli Hilmarsson

Fork-tailed Flycatcher *Tyrannus savana*

Breeds in Central and South America, from southern Mexico south to Brazil. Some southern populations migrate north and occur as vagrants in North America (Gutiérrez 2008; del Hoyo *et al.* 2004).

The occurrence of vagrants on the north-eastern coast of North America prompted the Spanish rarities committee to accept this as a genuine record of a vagrant (cf. Gutiérrez 2008).

1 record

Spain (1)
1 • 19 October 2002, Almonte, marismas de El Rocío, Puente de Ajolí, Huelva [Díes *et al.* 2007; Gutiérrez 2008]

VIREOS Vireonidae (n=3)

White-eyed Vireo *Vireo griseus*

Breeds in North America, from Iowa east to Massachusetts and south to Texas, Mexico and Florida. Winters in southern part of range and Central America and Caribbean (del Hoyo *et al.* 2010).

The second and third records for the Azores occurred in October 2008 and October 2009 (Gantlett 2009, 2010).

1 record

Azores (1)
1 • 22–23 October and 23 November 2005, Ribeira da Ponte, Corvo, first-winter [Jara *et al.* 2008]

22 October 2005, Ribeira da Ponte, Corvo, Azores. © Peter Alfrey

Yellow-throated Vireo *Vireo flavifrons*

Breeds in North America, from southern Manitoba and Minnesota east to Maine and south to eastern Texas, coast of Gulf of Mexico and central Florida. Winters in Central America, from southern Mexico to Colombia, Venezuela, Cuba and Bahamas (Cramp & Perrins 1994a).

The first three records for the Azores occurred in October 2008 (two) and October 2009 (Gantlett 2009, 2010).

2 records

Britain (1)
1 • 20–27 September 1990, Kenidjack, Cornwall [Birch 1990, 1994; British Ornithologists' Union 1992b; Rogers & Rarities Committee 1992]

Germany (1)
1 • 18 September 1998, Helgoland, Schleswig-Holstein [Aumüller 2005; Deutsche Seltenheitenkommission 2002]

September 1990, Kenidjack, Cornwall, Britain. © Tim Loseby

Extremely rare birds in the Western Palearctic

Philadelphia Vireo *Vireo philadelphicus*

Breeds in North America, from northeastern British Columbia, east to central Quebec and Newfoundland, south to North Dakota, northern Michigan and central Maine. Winters in Central America, from Yucatán peninsula to Panama and northern Colombia (Cramp & Perrins 1994a).

All records are from October, which is typical for a Nearctic vagrant that migrates along the Atlantic coast of North America. The first and second for the Azores were reported in October 2005 and October 2009 (Gantlett 2010; João Jara in litt).

3 records

Britain (1)
1 • 10–13 October 1987, Tresco, Isles of Scilly [British Ornithologists' Union 2003; Brodie Good 1991; Brodie Good & Filby 1987; Rogers & Rarities Committee 1991]

Ireland (2)
1 • 12–17 October 1985, Galley Head, Cork [Brazier *et al.* 1986; British Ornithologists' Union 1988; Dowdall 1995]
2 • 13–14 October 2008, Kilbaha, Loop Head, Clare [Paul Milne in litt]

October 2008, Kilbaha, near Loop Head, Clare, Ireland. © Chris Batty

October 1985, Galley Head, Cork, Ireland. © Richard Mills

SPECIES ACCOUNTS

SHRIKES Laniidae (n=1)

Long-tailed Shrike *Lanius schach*

Breeds from Kazakhstan and Turkmenistan east through India, China, Taiwan, Philippines, Indochina, Greater and Lesser Sundas to New Guinea. Northern populations winter in India, Burma and Indochina (Cramp & Perrins 1993; del Hoyo *et al.* 2008).

A record from Hungary in April 1979 (Cramp & Perrins 1993; Lewington *et al.* 1991) turned out to be a hybrid between Red-backed *L. collurio* and Woodchat Shrikes *L. senator* (cf. Hadarics & Schmidt 1997).

November 2000, Howbeg and Howmore, South Uist, Outer Hebrides, Britain. © Steve Gantlett

24 September 1987, Birecik, Turkey. © Natural History Museum

Long-tailed Shrike *Lanius schach*

8 records

Britain (1)
1 • 27 October to 04 November 2000, Howbeg and Howmore, South Uist, Outer Hebrides, first-winter [British Ornithologists' Union 2005; Rogers & Rarities Committee 2004; Stevenson 2000, 2005]

Denmark (1)
1 • 15–17 October 2007, Skallingen, Ribe, first-winter male [Kristensen *et al.* 2008]

Israel (1)
1 • 26 January to late February 1983, Sede Boqer, adult male [Shirihai & Golan 1994]

Jordan (1)
1 • 11–13 April 2004, Aqaba [Dufourny 2006]

Kuwait (2)
1 • 07 October 2004, Jahra Pool Reserve, Al Jahra [Foster 2006; KORC 2008]
2 • 14 October 2006 to 9 April 2007, Sulaibikhat Nature Reserve, Al Asimah [KORC 2008]
• 03 November 2007 to 12 April 2008, Sulaibikhat Nature Reserve, Al Asimah

Sweden (1)
1 • 11 June 1999, Stora Karlsö, Gotland, second-year male [Cederroth 2000]

Turkey (1)
1 • 24 September 1987, Birecik, first-winter male, trapped, collected (NHM: 1988.23.1) [Kirwan & Martins 1994; Shirihai & Golan 1994]

October 2007, Skallingen, Ribe, Denmark.
© Ole Krogh

SPECIES ACCOUNTS

11 June 1999, Stora Karlsö, Gotland, Sweden. © Peter Bergman

7 October 2004, Jahra Pool Reserve, Al Jahra, Kuwait. © Khaled Al-Ghanem

April 2004, Aqaba, Jordan. © Guy Conrady

3 November 2006, Sulaibikhat Nature Reserve, Al Asimah, Kuwait. © Mike Pope

February 1983, Sede Boqer, Israel. © Y. Golan

131

CROWS Corvidae (n=2)

Daurian Jackdaw *Corvus dauuricus*

Breeds from c 96° E in southern Siberia east to Amurland and Ussuriland, south through Mongolia and Manchuria to northern and western China. Winters south to Turkestan, Korea, Japan and southern China (Cramp & Perrins 1994a; del Hoyo *et al.* 2009).

The records from the Netherlands in 1995 and 1997 might relate to the same wandering bird given the period (both occurred in early May) and the same route north to south along the coast. The Danish record in 1997 and the French record in 1995 are most likely to relate to the same bird as the Dutch record. Two records from Germany in December 1995 and June 2005 were placed in category D (Deutsche Seltenheitenkommission 1997, 2008).

Early May 1883, Uusikaarlepyy, Finland. © Mikko Heikkinen

April 1985, Röbäcksslätten, Umeå, Västerbotten, Sweden. © Björn Olsen

8 May 1997, Balgzanddijk, Anna Paulowna, Noord-Holland, Netherlands. © Ruud Brouwer

SPECIES ACCOUNTS

4–6 records

Denmark (1)

1 • 12 April 1997, Blavåndshuk, Ribe, adult [Rasmussen 1998] [2]

Finland (1)

1 • early May 1883, Uusikaarlepyy, adult, collected (FMNH: AVE-FIN 2697) [Rariteettikomitea 1984; Stjernberg 1999]

France (1)

1 • 22 June 1995, Noirmoutier et La Guérinière, Vendée, adult [Dubois & CHN 1997; Dubois *et al.* 2008; Fouquet 1996] [1]

Netherlands (2)

1 • 01 May 1995, Hargen, Bergen, Noord-Holland, adult [van den Berg & Bosman 2001; Meijer 1996; Wiegant *et al.* 1997] [1]
• 04–06 May 1995, Wantveld, Katwijk aan Zee, Katwijk, Zuid-Holland, adult
• 07 May 1995, Ganzenhoek, Wassenaar, Zuid-Holland, adult
• 07 May 1995, Westduinpark, Den Haag, Zuid-Holland, adult
• 13–15 May 1995, Scheveningen, Den Haag, Zuid-Holland, adult

2 • 08 May 1997, Amsteldiepdijk, Wieringen, Noord-Holland & Balgzanddijk, Anna Paulowna, Noord-Holland, adult [van den Berg & Bosman 2001; Brouwer & Halff 1999; Wiegant *et al.* 1999] [2]
• 09–10 May 1997, Donkere Duinen, Den Helder, Noord-Holland, adult
• 10–11 May 1997, Egmond aan Zee, Bergen, Noord-Holland, adult

Sweden (1)

1 • 26–28 April 1985, Röbäcksslätten, Umeå, Västerbotten [Delin 1987; Olsson 1988]

[1] Possibly the same individual
[2] Possibly the same individual

May 1995, Katwijk aan Zee, Zuid-Holland, Netherlands. © René van Rossum

133

Pied Crow *Corvus albus*

Breeds in sub-Saharan Africa, Madagascar, Comoros and Aldabra group in western Indian Ocean (Cramp & Perrins 1994a; del Hoyo *et al.* 2009). Mainly a resident bird.

Recent records from Spain and the Canary Islands probably relate to escaped birds and have been placed in category D. Three birds seen in Western Sahara from December 2009 into 2010 were found breeding there in April 2010 (van den Berg & Haas 2010c; Gantlett 2010) and constitute the first record for Morocco. The first record for Egypt occurred in April 2010 (van den Berg & Haas 2010c).

3 records

Algeria (2)
1 • 1961, no location [Dupuy 1969; Isenmann & Moali 2000]
2 • December 1964, In Azoua [Dupuy 1969; Isenmann & Moali 2000]

Libya (1)
1 • 24 April 1931, Jalo oasis, Al Wahat, adult male, collected (MCGD: CE 31824) [Ghigi 1932]

24 April 1931, Jalo oasis, Al Wahat, Libya. © Enrico Borgo

SPECIES ACCOUNTS

GOLDCRESTS
Regulidae (n=1)

Ruby-crowned Kinglet *Regulus calendula*

Breeds in North America, from Alaska to Labrador and Newfoundland south to southern California, Guadeloupe Island (Mexico), Arizona, New Mexico, northern Michigan, Maine and Nova Scotia. Winters from southern British Columbia, Nebraska, southern Ontario and New Jersey south to Florida and western Guatemala (Cramp 1992; del Hoyo *et al.* 2006). It was recorded in Greenland in 1860, October 1956 (two) and September 1993 (Boertmann 1994; David Boertmann in litt).

A British record of a bird shot in summer of 1852 was not accepted because of its uncertain origin (Anonymous 1992; British Ornithologists' Union 1984, 1993b).

2 records

Iceland (2)

1 • 23 November 1987, Heimaey, Vestmannaeyjar, immature, collected (IMNH: RM9957) [Petersen 1989; Pétursson & Ólafsson 1989]

2 • 10–11 October 1998, Heimaey, Vestmannaeyjar, female or first-winter [Kolbeinsson *et al.* 2001]

10 October 1998, Heimaey, Vestmannaeyjar, Iceland. © Örn Óskarsson

23 November 1987, Heimaey, Vestmannaeyjar, Iceland. © Yann Kolbeinsson

135

LARKS Alaudidae (n=2)

Chestnut-headed Sparrow-Lark *Eremopterix signatus*

Breeds in Africa, from Ethiopia and Somalia south to south-eastern Sudan and Kenya (del Hoyo *et al.* 2004).

This bird was probably an overshooting migrant.

1 record

Israel (1)
1 • 01 May 1983, Eilot, adult male [Shirihai 1996, 1999]

Hume's Short-toed Lark *Calandrella acutirostris*

Breeds in Asia, from north-eastern Iran east to China and south to northern India. Winters in Pakistan and northern India (del Hoyo *et al.* 2004).

The identification of this bird is discussed by Shirihai & Alström (1990).

1 record

Israel (1)
1 • 04–14 February 1986, Eilot, trapped (14 February), died (TAU: Av.9983) [Shirihai 1996, 1999]

February 1986, Eilot, Israel. © Hadoram Shirihai

SWALLOWS
Hirundinidae (n=4)

Banded Martin *Riparia cincta*

Breeds in Africa, from Sudan and Ethiopia south to South Africa. Winters in Cameroon and Angola (del Hoyo *et al.* 2004).

The only record probably relates to a bird from either the Sudanese or Ethiopian populations.

1 record

Egypt (1)
1 • 15 November 1988, Elephantine Island, Aswan [Clements 1990]

Extremely rare birds in the Western Palearctic

Tree Swallow *Tachycineta bicolor*

Breeds in North America, from Alaska and Canada south to California, Nevada, Arizona, New Mexico, Texas, Alabama, Georgia and North Carolina. Winters mainly along coast of southern North America and West Indies to northern coast of South America (del Hoyo *et al.* 2004). It was recorded in Greenland in July 1864, April 1941, May 1948, May 1958, June 1970 and June 1977, and once without any known details (Boertmann 1994).

An additional four records occurred in the Azores in October 2005, November 2005 (two records involving three birds) and October 2007; the corresponding descriptions have not yet been received or accepted by the rarities committee (João Jara in litt).

2 November 2005, Sete Cidades, São Miguel, Azores. © Jens Hering

29 May 2002, Burrafirth, Unst, Shetland, Britain. © Wendy Dickson

4 records (5 individuals)

Azores (2/3)
1 • 02 November 2005, Sete Cidades, São Miguel, *two* [Jara *et al.* 2007]
2 • 19–20 October 2007, Corvo [João Jara in litt]

Britain (2)
1 • 06–10 June 1990, St Mary's, Isles of Scilly [British Ornithologists' Union 1992b; Hickman 1995; Rogers & Rarities Committee 1992; Wagstaff 1990]
2 • 29 May 2002, Burrafirth, Unst, Shetland [Rogers & Rarities Committee 2003]

SPECIES ACCOUNTS

19 October 2007,
Corvo, Azores.
© Vincent Legrand

June 1990, St Mary's,
Isles of Scilly, Britain.
© David Cottridge

Purple Martin *Progne subis*

Breeds in North America, from southern Canada south to Mexico. Winters in South America east of Andes (del Hoyo *et al.* 2004).

Remarkably, the two records occurred within a day of each other. A record of a bird collected *c*. March 1840 in Dublin, Ireland, has been placed in category D (Irish Rare Birds Committee 1998). Clarke (2006) mentions a record of two birds on Corvo, Azores, in September 1996, which is now under review by the Portuguese rarities committee (João Jara in litt).

2 records

Azores (1)
1 • 06 September 2004, Facho, Flores, first-winter [Coyle *et al.* 2004; Jara *et al.* 2007]

Britain (1)
1 • 05–06 September 2004, Butt of Lewis, Outer Hebrides, immature [British Ornithologists' Union 2006a; Coyle *et al.* 2004, 2007; Rogers & Rarities Committee 2005]

6 September 2004, Facho, Flores, Azores. © Ingvar Torsson

September 2004, Butt of Lewis, Outer Hebrides, Britain. © Martin Scott

Ethiopian Swallow *Hirundo aethiopica*

Breeds in Africa, from Senegal and Gambia east to Sudan, Ethiopia, Somalia and Kenya and south to Cameroon, Central African Republic and northern Tanzania (del Hoyo *et al.* 2004).

A record from Egypt in March 2004 will be reviewed by the Egyptian rarities committee in 2011 (Frédéric Jiguet in litt).

1 record

Israel (1)
1 •22 March 1991, Bet She'an, adult, trapped, released (Nir David, 23 March) [Bear 1991; Yoav Perlman in litt; Shirihai 1996]

23 March 1991, Bet She'an, Israel. © Yoav Perlman

LEAF WARBLERS
Phylloscopidae
(n=2)

Eastern Crowned Warbler *Phylloscopus coronatus*

Breeds in eastern Siberia, from Argun' river along Amur river to its mouth and south to western Manchuria, central and south-eastern Szechwan (China), Korea and Honshu (Japan). Winters in Assam, Bangladesh, Burma, Thailand, Indochina, Malaya, Sumatra and Java (Cramp 1992; del Hoyo *et al.* 2006).

The first record for Britain occurred in October 2009 (British Ornithologists' Union 2011, Holden & Bilton 2009; Hudson & Rarities Committee 2010).

4 records

Finland (1)
1 • 23 October 2004, Harrbåda, Kokkola [Luoto *et al.* 2005]

Germany (1)
1 • 04 October 1843, Helgoland, Schleswig-Holstein, collected (skin lost) [Gätke 1900]

Netherlands (1)
1 • 05 October 2007, Katwijk aan Zee, Katwijk, Zuid-Holland, first-year [Ovaa *et al.* 2008, Zuyderduyn 2008]

Norway (1)
1 • 30 September 2002, Jæren, Rogaland, first-winter, trapped [Bunes & Solbakken 2004]

30 September 2002, Jæren, Rogaland, Norway.
© Fredrik Kræmer

SPECIES ACCOUNTS

23 October 2004, Harrbåda, Kokkola, Finland. © Harri Taavetti

5 October 2007, Katwijk aan Zee, Katwijk, Zuid-Holland, Netherlands. © René van Rossum

143

Plain Leaf Warbler *Phylloscopus neglectus*

Breeds in Asia, in mountains of Iran, southern Turkmenistan, Tajikistan, Afghanistan, northern Baluchistan and Kashmir. Winters south to Oman, coastal Iran, western Punjab and Sind (Cramp 1992; del Hoyo *et al.* 2006).

A record from Jordan in April 1963 is not accepted (cf. Ian Andrews in litt; Cramp 1992). Kirwan *et al.* (2008) mention recent reports from Turkey in June 2004 and May–June 2005, but these have not been accepted; however, they will be reviewed. Ramadan-Jaradi *et al.* (2008) mention a record of four birds in October 1996 in Lebanon, which will be reviewed by the Lebanese rarities committee (Ghassan Ramadan-Jaradi in litt).

1 record

Sweden (1)
1. • 10 October 1991, Landsort, Södermanland, trapped [Elmberg 1992]

SPECIES ACCOUNTS

GRASSHOPPER WARBLERS
Locustellidae (n=1)

Gray's Grasshopper Warbler *Locustella fasciolata*

Breeds in central and eastern Siberia north to *c.* 60° N from Novosibirsk east to Sakhalin, Kuril Islands, northern Manchuria and Korea. Winters in Philippines, Sulawesi, Moluccas and western New Guinea (Cramp 1992; del Hoyo *et al.* 2006).

A report of a bird on Ouessant, Finistère, France, in September 1933 (Cramp 1992; Lewington *et al.* 1991) is not accepted (cf. CAF 2006; Dubois *et al.* 2008; Kennerley & Prŷs-Jones 2006).

2 records

Denmark (1)
1 • 25 September 1955, Lodbjerg Fyr, Nordjylland, first-winter female, found dead (ZMUC: Tv.J: 23-2-1960-24) [Dybbro 1978]

France (1)
1 • 26 September 1913, Creac'h, Ouessant, Finistère, first-winter female, found dead (NHM: 1929.10.7.1) [Dubois *et al.* 2008; Ingram 1929, 1930]

25 September 1955, Lodbjerg Fyr, Nordjylland, Denmark. © Troels Eske Ortvad

26 September 1913, Creac'h, Ouessant, Finistère, France. © Natural History Museum

145

REED WARBLERS
Acrocephalidae
(n=2)

Thick-billed Warbler *Iduna aedon*

Breeds in southern Siberia, from Ob to northern Mongolia east to Ussuriland, Manchuria and Hopeh. Winters in southern China, Indochina, India and Burma (Cramp 1992; del Hoyo *et al.* 2006). It has also been recorded as a vagrant in Oman (Eriksen *et al.* 2010).

Seven of the eight records occurred in September–October and there is just one spring record. Most records are from northern Europe; the Egyptian record is quite remarkable, although Eastern Palearctic warblers have occurred in North Africa and the Middle East before (e.g. Oriental Reed Warbler).

8 records

Britain (4)
1. • 06 October 1955, Fair Isle, Shetland, trapped [British Ornithologists' Union 1958; Williamson *et al.* 1956]
2. • 23 September 1971, Whalsay, Shetland, trapped, released (Lerwick, Shetland; 24 September 1971), killed by a cat (25 September 1971) (NMS: Z.1989.31) [Smith & Rarities Committee 1972]
3. • 14 September 2001, Out Skerries, Shetland, first-winter, trapped [Harvey 2001; Rogers & Rarities Committee 2002]

May 2003, Fair Isle, Shetland, Britain.
© Hugh Harrop

SPECIES ACCOUNTS

4 • 16–17 May 2003, Fair Isle, Shetland, trapped [Rogers & Rarities Committee 2004; Shaw 2003a]

Egypt (1)
1 • 20 November 1991, St Catharine monastery, Sinai [Grieve 1992]

Finland (1)
1 • 11 October 1994, Västra Norrskär, Mustasaari, trapped [Eischer 1995; Jännes 1995; Nikander & Lindroos 1995]

Norway (2)
1 • 06 and 08 October 2004, Utsira, Rogaland, trapped (06 October) [Mjølsnes *et al.* 2006]
2 • 03 October 2005, Utsira, Rogaland, trapped [Olsen & Mjølsnes 2007]

25 September 1971, Lerwick, Shetland, Britain.
© National Museums Scotland

6 October 2004, Utsira, Rogaland, Norway.
© Atle Grimsby

Oriental Reed Warbler *Acrocephalus orientalis*

Breeds in Asia, from eastern Sinkiang, eastern Mongolia and south-western Transbaykalia east to lower Amur river and Japan south through eastern China to Fukien. Winters in north-eastern India, Burma and throughout south-eastern Asia (Cramp 1992; del Hoyo *et al.* 2006).

A record from Kuwait in April 2004 mentioned by Gregory (2005) has not been accepted (Mike Pope in litt).

2 records

Israel (2)

1 • 28 February to 13 April 1988, Eilat [Shirihai 1996, 1999]
2 • 02 May 1990, Eilat, trapped [Shirihai 1996, 1999]

SPECIES ACCOUNTS

WAXWINGS
Bombycillidae (n=1)

Cedar Waxwing *Bombycilla cedrorum*

Breeds in North America, from Alaska east through southern Canada to Newfoundland and south to California and Georgia. Winters south to Central America (del Hoyo *et al.* 2005).

The 1996 British record involved a bird within a large group of Bohemian Waxwings *B. garrulous*; the other records are of single birds. The first for Ireland occurred in October 2009 (McGeehan & Nash 2009).

April to July 1989, Gerðar, Gullbringusýsla, Iceland. © Þórhallur Frímannsson

4 records

Britain (2)
1. • 25–26 June 1985, Noss, Shetland [British Ornithologists' Union 1993a, 1998; McKay 2000; Rogers & Rarities Committee 1993]
2. • 20 February to 18 March 1996, Nottingham, Nottinghamshire, first-winter [British Ornithologists' Union 1998; Rogers & Rarities Committee 1997; Smith 1996]

Iceland (2)
1. • mid-April to late July 1989, Gerðar, Gullbringusýsla [Pétursson 1995; Þráinsson *et al.* 1995]
2. • 08 October 2003, Heimaey, Vestmannaeyjar [Kolbeinsson *et al.* 2006]

Extremely rare birds in the Western Palearctic

Cedar Waxwing *Bombycilla cedrorum*

June 1985, Noss, Shetland, Britain. © Clive McKay

February 1996, Nottingham, Nottinghamshire, Britain. © Steve Young

8 October 2003, Heimaey, Vestmannaeyjar, Iceland. © Yann Kolbeinsson

SPECIES ACCOUNTS

NUTHATCHES
Sittidae (n=1)

Red-breasted Nuthatch *Sitta canadensis*

Breeds in southern Canada and USA to California and Texas in the west and Appalachian mountains in the east. Winters in southern part of breeding range to southern USA (Cramp & Perrins 1993; del Hoyo *et al.* 2008).

The British bird was seen by more than 5000 people (cf. Evans 1994).

2 records

Britain (1)

1 • 13 October 1989 to at least 06 May 1990, Holkham Meals, Norfolk, first-year male [Aley & Aley 1995; British Ornithologists' Union 1991b; Hatton & Varney 1989; Rogers & Rarities Committee 1991, 1996]

Iceland (1)

1 • 21 May 1970, Heimaey, Vestmannaeyjar, adult male, found dead (IMNH: RM5317) [Pétursson & Þráinsson 1999]

January 1990, Holkham Meals, Norfolk, Britain. © Alan Tate

21 May 1970, Heimaey, Vestmannaeyjar, Iceland. © Yann Kolbeinsson

MOCKINGBIRDS
Mimidae (n=3)

Northern Mockingbird *Mimus polyglottos*

Breeds in North America, from southern British Columbia east to Nova Scotia and south to southern Mexico. Mainly resident but some northern populations move south in winter (del Hoyo *et al.* 2005).

Two additional records in Britain (August 1971 and July–August 1978) are considered unlikely to have been genuine vagrants (British Ornithologists' Union 1980, 1993b). A record from Gran Canaria, Canary Islands, from November 2004 to January 2006 has been placed in category D (de Juana & Comité de Rarezas de la SEO 2006).

3 records

Britain (2)

1 • 30 August 1982, Saltash, Cornwall [British Ornithologists' Union 1993b; Griffiths 1996; Rogers & Rarities Committee 1994]

2 • 17–23 May 1988, Hamford Water, Essex [British Ornithologists' Union 1993b; Cox 1988, 1996; Rogers & Rarities Committee 1994]

Netherlands (1)

1 • 16–23 October 1988, Schiermonnikoog, Friesland [van den Berg & Bosman 2001; van den Berg *et al.* 1991; de By & CDNA 1991; Ebels 1991]

SPECIES ACCOUNTS

May 1988, Hamford Water, Essex, Britain. © Pete Loud

October 1988, Schiermonnikoog, Friesland, Netherlands. © Sander van de Water

Brown Thrasher *Toxostoma rufum*

Breeds in central and eastern North America, from southern Canada to eastern Texas and Florida. Winters in eastern and southern North America (Cramp 1988; del Hoyo *et al.* 2005).

A bird collected in the autumn of 1836 on Helgoland, Schleswig-Holstein, Germany, is not accepted because the specimen cannot be traced (cf. Barthel & Helbig 2005).

1 record

Britain (1)
1 • 18 November 1966 to 05 February 1967, Durlston Head, Dorset, trapped (23 November) [British Ornithologists' Union 1971a; Incledon 1968; Smith & Rarities Committee 1967, 1968]

23 November 1966, Durlston Head, Dorset, Britain. © David Godfrey

Grey Catbird *Dumetella carolinensis*

Breeds in North America, from southern Canada, eastern Oregon, central Arizona and north-eastern Texas south to central part of Gulf states. Winters in eastern North America, from New England south to Panama and Caribbean Islands (Cramp 1988; del Hoyo *et al.* 2005).

In October 1998, a bird stayed on board a ship from New York, USA, as far as the Mediterranean but, because it had been fed during its journey, the record was not accepted by the rarities committees of the countries involved (Anonymous 1998a).

7 records

Belgium (1)
1 • 15–16 December 2006, Kallo, Oost-Vlaanderen, first-winter [Symens & Spanoghe 2007; Vandegehuchte & BAHC 2008]

Britain (1)
1 • 04–06 October 2001, South Stack, Anglesey [British Ornithologists' Union 2003; Croft 2001, 2004; Rogers & Rarities Committee 2002, 2005]

Canary Islands (1)
1 • 04 November 1999, La Mareta, Tenerife [de Juana 2006; de Juana & Comité de Rarezas de la SEO 2004]

Channel Islands (1)
1 • mid-October 1975, St Brelade, Jersey, trapped and kept in captivity until it died a few years later [Long 1981ab; Roger Long in litt]

Germany (2)
1 • 28 October 1840, Helgoland, Schleswig-Holstein, collected (IVVH: C1124) [Gätke 1900]
2 • 02 May 1908, Leopoldshagen, Mecklenburg-Vorpommern [Heinroth 1908, Thiede 1966]

Ireland (1)
1 • 04 November 1986, Cape Clear Island, Cork [British Ornithologists' Union 1991a; Preston 1989; O'Sullivan & Smiddy 1987]

Extremely rare birds in the Western Palearctic

Grey Catbird *Dumetella carolinensis*

16 December 2006, Kallo, Oost-Vlaanderen, Belgium. © Vincent Legrand

28 October 1840, Helgoland, Schleswig-Holstein, Germany. © Jochen Dierschke

October 1975, St Brelade, Jersey, Channel Islands. © Roger Long

STARLINGS
Sturnidae (n=1)

Daurian Starling *Agropsar sturninus*

Breeds across central and eastern Asia, Transbaykalia, Amurland, Ussuriland, Mongolia, Manchuria, northern Korea and northern China. Winters from southern China south to southern Burma, Malaya, Sumatra and Java (Cramp & Perrins 1994a; del Hoyo *et al.* 2009).

All records in Britain have been placed in category D (British Ornithologists' Union 1991a, 1994b). Other records from the Netherlands are also thought to relate to escapes rather than genuine vagrants (cf. Ebels 2004).

2 records

Netherlands (1)
1 • 11–12 October 2005, Oost-Vlieland, Vlieland, Friesland, first-winter male [van der Vliet *et al.* 2007]

Norway (1)
1 • 29 September 1985, Lillestrøm, Akershus, first-year, collected (specimen lost) [Bentz 1987; Jan T Lifjeld in litt; Viggo Ree in litt]

12 October 2005, Oost-Vlieland, Vlieland, Friesland, Netherlands. © Jan van der Laan

Extremely rare birds in the Western Palearctic

Daurian Starling *Agropsar sturninus*

29 September 1985, Lillestrøm, Akershus, Norway. © Viggo Ree

SUNBIRDS
Nectariniidae (n=1)

Purple Sunbird *Cinnyris asiaticus*

Breeds from UAE, northern Oman and south-eastern Iran east through India to northern Thailand, Cambodia and Vietnam. Mainly resident but some dispersive movements occur (Cheke *et al.* 2001; del Hoyo *et al.* 2008).

This record probably concerns birds from either the Arab population or from Iran (cf. Al Hajji *et al.* 2008).

1 record (3 individuals)

Kuwait (1/3)

1 • 05 January to 10 February 2008, Ras Al Subiyah, *maximum of three* (males) [Al Hajji *et al.* 2008; Mike Pope *in litt*]

11 January 2008, Ras Al Subiyah, Kuwait. © Mike Pope

Extremely rare birds in the Western Palearctic

FLYCATCHERS
Muscicapidae (n=11)

Varied Thrush *Ixoreus naevius*

Breeds from Alaska to south-western Alberta, northern Idaho and north-western California. Winters mostly within its breeding range, east to Montana and southern Baja California (Clement & Hathway 2000; del Hoyo *et al.* 2005).

The British record was of an extremely rare aberrant form, lacking the usual orange pigmentation (Madge *et al.* 1990).

2 records

Britain (1)
1 • 14–23 November 1982, Nanquidno, Cornwall [British Ornithologists' Union 1991a; Madge *et al.* 1990; Rogers & Rarities Committee 1989]

Iceland (1)
1 • 03–08 May 2004, Unaós í Hjaltastaðaþinghá, Norður-Múlasýsla, male [Kolbeinsson 2004; Kolbeinsson *et al.* 2007; Þórisson 2007]

LEFT: November 1982, Nanquidno, Cornwall, Britain. © Tony Croucher

BELOW: 3 May 2004, Unaós í Hjaltastaðaþinghá, Norður-Múlasýsla, Iceland. © Skarphéðinn Þórisson

Wood Thrush *Hylocichla mustelina*

Breeds in extreme south-eastern Canada and eastern USA, from south-eastern Ontario and south-western Quebec east to New Brunswick and south-western Nova Scotia. Winters from central Mexico to Panama (Clement & Hathway 2000; del Hoyo *et al*. 2005).

The date of the Azores record cannot be determined exactly but the bird was probably collected in autumn like the other two records.

3 records

Azores (1)
1 • before 1900, Ponta Delgada, São Miguel, collected (MCM: MCM1010) [Bannerman & Bannerman 1966]

Britain (1)
1 • 07 October 1987, St Agnes, Isles of Scilly, first-winter [British Ornithologists' Union 1991b; Dukes 1987, 1995; Rogers & Rarities Committee 1991]

Iceland (1)
1 • 23 October 1967, Kvísker í Öræfum, Austur-Skaftafellssýsla, male, collected (IMNH: RM4746) [Pétursson & Þráinsson 1999]

Before 1900, Ponta Delgada, São Miguel, Azores. © Museu Carlos Machado / Staffan Rodebrand

23 October 1967, Kvísker í Öræfum, Austur-Skaftafellssýsla, Iceland. © Yann Kolbeinsson

Veery *Catharus fuscescens*

Breeds in North America, from interior British Columbia east to southern Newfoundland and south to north-eastern Arizona, South Dakota, Minnesota, Alleghenies, Georgia, New York and Long Island. Winters in northern South America (Clement & Hathway 2000; Cramp 1988; del Hoyo *et al.* 2005).

Seven of the eight records are from late September–October. The May record is remarkable, although there are other spring records of *Catharus* thrushes in western Europe. The eighth and ninth records for Britain occurred in October 2009 (Hudson & Rarities Committee 2010; Millington 2009).

6 October 1970, Porthgwarra, Cornwall, Britain. © Keith Allsopp

8 records

Britain (7)
1. •06 October 1970, Porthgwarra, Cornwall, trapped [Allsopp 1972; British Ornithologists' Union 1972; Smith & Rarities Committee 1972]
2. •10 October to 11 November 1987, Lundy, Devon, (re)trapped (10 and 20 October) [Aley 1987; King 1990; Rogers & Rarities Committee 1988;]
3. •20–28 October 1995, Newton, North Uist, Outer Hebrides [Rogers & Rarities Committee 1997, 1999]
4. •14 May 1997, Lundy, Devon, trapped [Campey 1997; Rogers & Rarities Committee 1999]
5. •13 October 1999, St Levan, Cornwall, first-winter [Rogers & Rarities Committee 2000]
6. •30 September to 05 October 2002, North Ronaldsay, Orkney, trapped (30 September) [Rogers & Rarities Committee 2003]
7. •22 September 2005, Northdale, Unst, Shetland, first-winter, trapped, later killed by a cat (skin not kept) [Fraser *et al.* 2007b]

Sweden (1)
1. •26 September 1978, Svenska Högarna, Uppland, trapped [Edholm *et al.* 1980; Ohlsson 1980]

SPECIES ACCOUNTS

October 1987, Lundy, Devon, Britain. © David Cottridge

13 October 1999, St Levan, Cornwall, Britain. © Rob Wilson

26 September 1978, Svenska Högarna, Uppland, Sweden. © Arne Lundberg

14 May 1997, Lundy, Devon, Britain. © Richard Campey

30 September 2002, North Ronaldsay, Orkney, Britain. © Ross McGregor

22 September 2005, Northdale, Unst, Shetland, Britain. © Micky Maher

Tickell's Thrush *Turdus unicolor*

Breeds in Himalayas, from northern Pakistan through Kashmir and northern India to Nepal, Sikkim and Bhutan. Winters at lower altitudes within breeding range but also in adjacent plains and foothills south and east of breeding range (Clement & Hathway 2000; del Hoyo *et al.* 2005).

The bird from Helgoland was shot and subsequently identified and represents a remarkable record of this short-distance migrant. The three white feathers on its right wing and belly are odd, but the bird showed no other signs of captivity and so there is no reason to doubt its wild origin.

1 record

Germany (1)
1 • 15 October 1932, Helgoland, Schleswig-Holstein, adult male, collected (IVVH: C3992) [Drost 1933]

15 October 1932, Helgoland, Schleswig-Holstein, Germany. © Jochen Dierschke

Asian Brown Flycatcher *Muscicapa dauurica*

Breeds in southern Siberia, from Yenisey east to Transbaykalia, northern Mongolia, Ussuriland, Manchuria, Sakhalin, northern Korea, Japan and Kuril Islands; also India. Winters in China, Philippines, Burma, Indochina, Malaysia, Sumatra, Java, Borneo, India and Sri Lanka (Cramp & Perrins 1993; del Hoyo *et al.* 2006).

A German record from Helgoland in August 1982 was re-identified as Dark-sided Flycatcher *M. sibirica* and has been accepted as an escape (Fleet 1982; Stühmer 2005).

1 July 1992, The Plantation, Fair Isle, Shetland, Britain. © Paul Harvey

6 records

Britain (3)
1 • 01–02 July 1992, The Plantation, Fair Isle, Shetland, trapped (01 July), first-summer [British Ornithologists' Union 1994a, 2001, 2010; Harvey 1992, 2010; Hudson & Rarities Committee 2009; Parkin & Shaw 1994; Rogers & Rarities Committee 1994]
2 • 03–04 October 2007, Flamborough Head, East Yorkshire, adult [Baines 2007; British Ornithologists' Union 2010; Hudson & Rarities Committee 2009]
3 • 24–25 September 2008, Ward Hill, Fair Isle, Shetland, first-winter [British Ornithologists' Union 2010; Hudson & Rarities Committee 2009]

Denmark (1)
1 • 24–25 September 1959, Blåvandshuk, Ribe [Christensen 1960]

Greece (1)
1 • 04 September 1993, Nesto delta, first-year [Handrinos & Akriotis 1996]

Sweden (1)
1 • 27 and 30 September 1986, Svenska Högarna, Uppland, first-year, trapped (27 September) [Douhan 1989; Hirschfeld 1987]

Extremely rare birds in the Western Palearctic

Asian Brown Flycatcher *Muscicapa dauurica*

4 October 2007, Flamborough Head, East Yorkshire, Britain. © Chris Batty

September 1986, Svenska Högarna, Uppland, Sweden. © Peter Bjurenstål

24 September 2008, Ward Hill, Fair Isle, Shetland, Britain. © Steve Minton

September 1959, Blåvandshuk, Ribe, Denmark. © Niels Hesselbjerg Christensen

Rufous-tailed Robin *Luscinia sibilans*

Breeds in Asia in central and eastern Siberia and north-eastern China. Winters in southern and south-eastern China and south-eastern Asia (del Hoyo *et al.* 2005).

The British record came after an invasion of White's Thrushes *Zoothera aurea* into Europe and the presence of a Chestnut-eared Bunting *Emberiza fucata* on Fair Isle the week before (cf. Slack 2009).

2 records

Britain (1)
1 • 23 October 2004, Fair Isle, Shetland, first-winter, trapped [British Ornithologists' Union 2006b; Fraser *et al.* 2007b; Shaw 2004b, 2006]

Poland (1)
1 • 30–31 December 2005, Fasty, Białystok, Podlaskie, first-winter [Grygoruk & Tumiel 2006ab; Komisja Faunistyczna 2006]

ABOVE: 23 October 2004, Fair Isle, Shetland, Britain. © Hugh Harrop

BELOW: 31 December 2005, Fasty, Białystok, Podlaskie, Poland. © Tomasz Kulakowski

Siberian Blue Robin *Luscinia cyane*

Breeds in southern Siberia, from Altai east to Amurland and Sakhalin south to Manchuria, Korea, Japan and northern China. Winters in south-eastern China west to Burma, Philippines, Borneo and Sumatra (Cramp 1988; del Hoyo *et al.* 2005).

All records are from October. The two records in 2000 are remarkable since the only previous record was in 1975.

4 records

Britain (2)
1. • 23 October 2000, Minsmere, Suffolk, female/first-winter [British Ornithologists' Union 2003; Foster 2000, 2006; Rogers & Rarities Committee 2002]
2. • 02 October 2001, North Ronaldsay, Orkney, first-winter male [Brown 2001; Rogers & Rarities Committee 2002]

Channel Islands (1)
1. • 27 October 1975, Banquette Valley, Sark, trapped [Long1981a; Rountree 1977]

Spain (1)
1. • 18 October 2000, Ebro delta, first-winter male, trapped [Bigas & Gutiérrez 2000; de Juana 2006; de Juana & Comité de Rarezas de la SEO 2004]

18 October 2000, Ebro Delta, Spain. © David Bigas

2 October 2001, North Ronaldsay, Orkney, Britain. © John Kinsley

SPECIES ACCOUNTS

27 October 1975, Banquette Valley, Sark, Britain. © Alan Marsden

Daurian Redstart *Phoenicurus auroreus*

Breeds from Siberia and Mongolia east to Amurland and south to northeastern India, China and Korea. Winters from eastern Himalayas east to Japan, Taiwan, south-eastern China and northern Indonesia (del Hoyo *et al.* 2005).

In April 1988, a male was seen and later found dead in Britain but was assumed to be an escaped bird since the bird belonged to the more sedentary southern subspecies *leucopterus* (James 1988; British Ornithologists' Union 1992a; Knox 1993a). A record of a male from Sweden in September 1997 was placed in category D (Cederroth 1998).

1 record

Russia (1)

1 • 18–30 September 2006, Pechoro-Ilychskiy reserve, Upper Pechora, male [Ryabitsev 2008; Slack 2009]

September 2006, Pechoro-Ilychskiy reserve, Upper Pechora, Russia. © Nikolay Neyfeld

Ant Chat *Myrmecocichla aethiops*

Breeds in Africa, from Senegal to Niger and Lake Chad and northern Nigeria to northern Cameroon and Sudan and in highlands of Kenya and northern Tanzania (Cramp 1988; del Hoyo *et al.* 2005).

The specimen could not be assigned to any particular subspecies (cf. Cramp 1988).

1 record

Chad (1)
1 • 1954, Yebbi Bou, Tibesti, collected (specimen location unknown) [Simon 1965]

Variable Wheatear *Oenanthe picata*

Breeds in Asia, from south-western Caspian and central Iran east to Tajikistan and northern and western Pakistan. Winters in Pakistan and northern India (Cramp 1988; del Hoyo *et al.* 2005).

Reports from Jordan and Syria (Cramp 1988) were of dark-morph Eastern Mourning Wheatears O. *lugens* (cf. Andrews 1994; Tye 1994). Ramadan-Jaradi *et al.* (2008) mention a record from Lebanon in December 2000 that will be reviewed by the Lebanese rarities committee (Ghassan Ramadan-Jaradi in litt).

1 record

Israel (1)
1 • 04 February 1986, Eilat, dark morph [Shirihai 1996, 1999]

Mugimaki Flycatcher *Ficedula mugimaki*

Breeds in Asia, in southern Siberia, northern Mongolia, north-eastern China, northern Korea and Sakhalin. Winters in south-eastern China, south-eastern Asia, Greater Sundas, Philippines and Sulawesi (del Hoyo *et al.* 2006).

A British record in November 1991 was subsequently placed in category D (cf. Parkin & Shaw 1994; Parrish 1991; Rogers & Rarities Committee 1994; British Ornithologists' Union 1994a).

The species was also recorded in Italy in October 1957 but this report is not accepted (Fracasso *et al.* 2009).

1 record

Russia (1)
1 • 02 August 2007, Neftekamsk, Bashkortostan, adult male [Fominykh 2007]

Extremely rare birds in the Western Palearctic

WAGTAILS
Motacillidae (n=2)

Forest Wagtail *Dendronanthus indicus*

Breeds in China, Korea and adjacent parts of south-eastern Russia. Winters in south-eastern Asia, Indonesia and India (Alström & Mild 2003; del Hoyo *et al.* 2004).

It has also been recorded from Oman and UAE (Eriksen *et al.* 2003, 2010; Pedersen & Aspinall 2010).

1 record

Kuwait (1)

1 • 10 November 2006, Al-Abraq Al-Khabari [BMAPT & KEPS 2007; Pope *et al.* 2006]

10 November 2006, Al-Abraq Al-Khabari, Kuwait. © Mike Pope

Amur Wagtail *Motacilla leucopsis*

Breeds in south-eastern Russia, south through Ussuriland, north-eastern, eastern and southern China, North and South Korea, south-western Japan and Taiwan. Winters in southern China, south through Taiwan, Hainan, most of Indochina, Myanmar, Andaman Islands, Bangladesh, northern and north-eastern India and Nepal (Alström & Mild 2003; del Hoyo *et al.* 2004). It has also been recorded from Oman in February 2005 (Addinall 2010).

The Norwegian bird was identified correctly after studying the photographs.

2 records

Britain (1)
1 • 05–06 April 2005, Seaham, Durham, male [Addinall 2005, 2010; British Ornithologists' Union 2010; Hudson & Rarities Committee 2009]

Norway (1)
1 • 01–02 November 2008, Kalleberg, Farsund, Vest-Agder, male [Gantlett 2010; Olsen *et al.* 2010]

6 April 2005, Seaham, Durham, Britain. © Chris Batty

November 2008, Kalleberg, Farsund, Vest-Agder, Norway. © Lars Bergersen

FINCHES
Fringillidae (n=1)

Evening Grosbeak *Hesperiphona vespertina*

Breeds in North America, from south-western and northern-central British Columbia east to Nova Scotia and south through Rocky Mountains to southern Mexico and east of mountains south to central Minnesota, southern Ontario, northern New York and Massachusetts. Winters throughout breeding range, sporadically south to lowland areas of southern USA (Cramp & Perrins 1994a).

A German record of a pair in March 2006 was not placed in category A because the birds were regarded as having escaped from captivity (Deutsche Seltenheitenkommission 2009).

4 records

Britain (2)
1. • 26 March 1969, St Kilda, Outer Hebrides, male [British Ornithologists' Union 1971b; Picozzi 1971; Smith & Rarities Committee 1970]
2. • 10–25 March 1980, Nethybridge, Badenoch and Strathspey, Highland, adult female [Rogers & Rarities Committee 1981]

Norway (2)
1. • 02–09 May 1973, Krosnes, Gressvik, Onsøy, Østfold, male [Andersen 1974; Ree 1976]
2. • 17–26 May 1975, Halten, Sør-Trøndelag, male, trapped [Ree 1976]

May 1975, Halten, Sør-Trøndelag, Norway.
© Harald Støen

SPECIES ACCOUNTS

26 March 1969, St Kilda, Outer Hebrides, Britain. © Nick Picozzi

May 1973, Krosnes, Gressvik, Onsøy, Østfold, Norway. © K. Møklegård

BUNTINGS
Emberizidae (n=28)

Summer Tanager *Piranga rubra*

Breeds in southern and southern-central USA from California east to southern New Jersey, south to northern-central Mexico. Winters from southern-central Mexico south through Central America to Ecuador, northern Bolivia and Amazonian Brazil (Cramp & Perrins 1994b).

The Azorean record coincided with a period during which many Nearctic vagrants were recorded in the Azores (cf. van den Berg & Haas 2006).

2 records

Azores (1)
1 • 26–28 October 2006, Corvo, first-winter [João Jara in litt]

Britain (1)
1 • 11–25 September 1957, Bardsey, Caernarfonshire, first-winter male, (re)trapped (11, 15 and 20 September) [British Ornithologists' Union 1963; Nisbet 1963a; Rogers & Rarities Committee 1999]

September 1957, Bardsey, Caernarfonshire, Britain. © Bill Condry

28 October 2006, Corvo, Azores. © Vincent Legrand

Dickcissel *Spiza americana*

Breeds in North America, from eastern Montana and south-eastern Saskatchewan east to Massachusetts, south to central Colorado, southern Texas, southern Louisiana, central Mississippi and South Carolina. Winters in Mexico, Central America and northern South America, from Michoacán south to northern Colombia, Venezuela and Guianas; locally also in small numbers in coastal lowlands of USA from southern New England to southern Texas (Cramp & Perrins 1994b).

The first record for the Azores was of two birds together in November 2009 (Gantlett 2010).

1 record

Norway (1)
1 • 29 July 1981, Måløy, Sogn og Fjordana, adult male [Michaelsen 1985]

Eastern Towhee *Pipilo erythrophthalmus*

Breeds in North America, from southern British Columbia east to Maine and south throughout USA to highlands of Mexico and central Guatemala. Winters within breeding range, but vacates extreme north (Byers *et al.* 1995; Cramp & Perrins 1994b).

The bird trapped on Lundy in June 1966 was assigned to one of the eastern subspecies (now split as 'Eastern Towhee'). A bird recorded at Spurn, East Yorkshire, Britain, from 5 September 1975 to 10 January 1976 and belonging to one of the western subspecies (now split as 'Spotted Towhee *P. maculatus*') was not accepted because of its suspected captive origin (cf. Cudworth 1979; O'Sullivan & Rarities Committee 1977; British Ornithologists' Union 1980).

1 record

Britain (1)
1 • 07 June 1966, Lundy, Devon, trapped [British Ornithologists' Union 1971a, 1980, 1997; Smith & Rarities Committee 1967; Waller 1970]

Lark Sparrow *Chondestes grammacus*

Breeds in much of North America east to Mississippi. Winters in Central America. Northern populations are migratory (Byers *et al.* 1995).

The two British records probably relate to ship-assisted birds. Charlton (1991) and the editorial comment provide more information on the possibilities that these were true vagrants.

2 records

Britain (2)

1. • 30 June to 08 July 1981, Landguard Point, Suffolk [British Ornithologists' Union 1984, 1993a; Charlton 1995; Rogers & Rarities Committee 1982, 1993]
2. • 15–17 May 1991, Waxham, Norfolk [Rogers & Rarities Committee 1993]

May 1991, Waxham, Norfolk, Britain. © Steve Young

Savannah Sparrow *Passerculus sandwichensis*

Breeds in Alaska, Canada and much of North America and Central America. Northern populations migrate south to southern North America and West Indies (Byers *et al.* 1995).

The April 1982 record involved the subspecies *princeps* ('Ipswich Sparrow'), which breeds only on Sable Island, Nova Scotia, Canada, and winters on the Atlantic coast of North America from Nova Scotia south to Georgia (cf. Broyd 1985). The other records involved the subspecies that breeds in north-eastern North America. The second and third records for the Azores occurred in October 2009 (Gantlett 2010).

4 records

Azores (1)
1 • 31 October 2002, Fajã Grande, Flores, first-winter [Elias *et al.* 2004]

Britain (3)
1 • 11–16 April 1982, Portland, Dorset [British Ornithologists' Union 1986; Broyd 1985; Rogers & Rarities Committee 1985]
2 • 30 September to 01 October 1987, Fair Isle, Shetland, trapped (30 September) [Aspinall 1987; Ellis & Riddiford 1992; Rogers & Rarities Committee 1988, 1989]
3 • 14–19 October 2003, Fair Isle, Shetland [Rogers & Rarities Committee 2004; Shaw 2003b]

31 October 2002, Fajã Grande, Flores, Azores.
© Kris de Rouck

Extremely rare birds in the Western Palearctic

Savannah Sparrow *Passerculus sandwichensis*

April 1982, Portland, Dorset, Britain.
© David Cottridge

1 October 1987, Fair Isle, Shetland, Britain.
© Tim Loseby

15 October 2003, Fair Isle, Shetland, Britain. © Hugh Harrop

Fox Sparrow *Passerella iliaca*

Breeds from Alaska east to Newfoundland and western North America. Winters across southern North America (Byers *et al.* 1995). It was also recorded in Greenland in October 1910 and c 1945 (Boertmann 1994).

A record from Italy (Genova, Liguria, December 1936) is not accepted. Two records from Germany (Mellum, Niedersachsen, May 1949; and Scharhörn, Hamburg, April 1977) have been placed in category E (Barthel & Helbig 2005).

2 records

Iceland (1)

1 • 05 November 1944, Laugarholt í Bæjarsveit, Borgarfjarðarsýsla, male, collected (IMNH: RM5572) [Ólafsson 1993b; Pétursson & Þráinsson 1999]

Northern Ireland (1)

1 • 03–04 June 1961, Copeland Bird Observatory, Down, (re)trapped (03–04 June) [Ruttledge 1962; Wilde 1962]

5 November 1944, Laugarholt í Bæjarsveit, Borgarfjarðarsýsla, Iceland. © Yann Kolbeinsson

Chestnut-eared Bunting *Emberiza fucata*

Breeds in western Himalayas, China, south-eastern Siberia, Mongolia and Japan. Winters in southern parts of range, north to Korea and southern Japan, and south to northern Indochina (Byers *et al.* 1995).

Within a week of this record, a Rufous-tailed Robin, which has similar breeding and wintering ranges (see above), also turned up on Fair Isle.

1 record

Britain (1)
1. • 15–20 October 2004, Fair Isle, Shetland, first-winter male [British Ornithologists' Union 2007b; Fraser *et al.* 2007b; Shaw 2004a, 2008]

18 October 2004, Fair Isle, Shetland, Britain. © Hugh Harrop

Yellow-browed Bunting *Emberiza chrysophrys*

Breeds in south-eastern Siberia, from Irkutsk east to Barguzin mountains and Stanovoy range. Winters in central and south-eastern China (Byers *et al.* 1995; Cramp & Perrins 1994b).

A record from Ukraine in January 1983 (Davydovich & Gorban 1990) is unacceptable because of insufficient documentation (Geert Groot Koerkamp in litt). A record from Germany in April 2004 has been placed in category D or E (Deutsche Seltenheitenkommission 2008). The first record for Sweden occurred in January–February 2009 (Gantlett 2010).

ABOVE: 19 October 1982, Schiermonnikoog, Friesland, Netherlands. © Jan van der Straaten

CENTRE: May 1998, Hoy, Orkney, Britain. © Julian Sykes

BELOW: October 1980, Fair Isle, Shetland, Britain. © Ian Robertson

October 1994, St Agnes, Isles of Scilly, Britain. © George Reszeter

183

Yellow-browed Bunting *Emberiza chrysophrys*

8 records

Belgium (1)
1. • 20 October 1966, Tongeren, Limburg, first-winter male, trapped, kept in captivity until winter of 1969/70 when it died (skin lost) [Dufourny 1997; De Smet 1996; De Smet *et al.* 1996]

Britain (5)
1. • 19 October 1975, Holkham Meals, Norfolk, immature/female [British Ornithologists' Union 1991a; Holman 1990; Rogers & Rarities Committee 1989]
2. • 12–23 October 1980, Fair Isle, Shetland, male, trapped (12 October) [British Ornithologists' Union 1984; Kitson & Robertson 1983; Rogers & Rarities Committee 1982]
3. • 22–23 September 1992, North Ronaldsay, Orkney [Donnelly 1993; Rogers & Rarities Committee 1993]
4. • 19–22 October 1994, St Agnes, Isles of Scilly [Rogers & Rarities Committee 1995; Wright 1994]
5. • 04–05 May 1998, Hoy, Orkney [Rogers & Rarities Committee 1999]

France (1)
1. • autumn of 1827, Lille, Nord, first-winter male, trapped, collected (MDNL: 252) [Jiguet 2007; Jiguet *et al.* 2007]

Netherlands (1)
1. • 19 October 1982, Schiermonnikoog, Friesland, first-winter male, trapped [Blankert *et al.* 1984; van den Berg & Bosman 2001; Scharringa & Winkelman 1984; Vonk & van IJzendoorn 1988]

Autumn 1827,
Lille, Nord, France.
© Frédéric Jiguet

Chestnut Bunting *Emberiza rutila*

Breeds in eastern Siberia, from north-western Irkutsk region east to Sea of Okhotsk, south to Baikal region and, probably, northern Mongolia and northern Manchuria. Winters from Assam east to south-eastern China, south to northern Burma, Thailand and northern Indochina (Byers *et al.* 1995; Cramp & Perrins 1994b).

There have been an additional eight records in Britain: July 1974, June 1985, June 1986 (two), September 1994, May–June 1998, May 2000 and September 2002, all of which are considered to relate to captive birds. In Britain this bunting is, therefore, placed in category E (cf. British Ornithologists' Union 1978, 1992a, 1998, 2009; Hudson & Rarities Committee 2009). Both September records need to be treated with caution, however, because they may well have been genuine vagrants (but the same could be true for all these records). Note that September 2002 was also the month of the first Finnish record. It was also recorded in France between 1845 and 1874, in October 1995 and in October 2009, and in Belgium in October 1928 and April 1974; in both countries, it has been placed in category D (Joris Elst in litt; Dubois *et al.* 2008; Gantlett 2010). The bird in Slovenia was aged by Grošelj (1988) as a first-winter male, but was reassessed from photographs as an adult female (Nils van Duivendijk and Viggo Ree in litt).

10 October 1987, Godovič, Slovenia. © Dare Šere

Chestnut Bunting *Emberiza rutila*

5 records

Finland (1)
1 • 30 September to 01 October 2002, Hanko, Uusikaupunki, first-winter male, trapped (30 September) [Luoto *et al.* 2003]

Malta (1)
1 • 12 November 1983, Lunzjata, Gozo, first-winter male, trapped [Sultana & Gauci 1984–85]

Netherlands (1)
1 • 05 November 1937, Vogelringstation Wassenaar, Meijendel, Wassenaar, Zuid-Holland, first-winter female, collected (NNM: RMNH. AVES.149141) [van den Berg & Bosman 2001; van IJzendoorn *et al.* 1996; Junge & Koch 1938]

Norway (1)
1 • 13–15 October 1974, Utsira, Rogaland, first-winter male, trapped (14 October) (released on 15 October) [Ree 1976]

Slovenia (1)
1 • 10 October 1987, Godovi, adult female, trapped [Grošelj 1988]

5 November 1937, Vogelringstation Wassenaar, Meijendel, Wassenaar, Zuid-Holland, Netherlands. © NCB Naturalis, Leiden

12 November 1983, Lunzjata, Gozo, Malta. © Joe Sultana

SPECIES ACCOUNTS

14 October 1974, Utsira, Rogaland, Norway. © Viggo Ree

30 September 2002, Hanko, Uusikaupunki, Finland. © Pekka Alho

Brown-headed Cowbird *Molothrus ater*

Breeds in North America, from north-eastern British Columbia east to central Quebec and Novia Scotia, south to Mexico. Winters in southern and eastern North America, north to northern-central California, Texas, Michigan, southern Ontario, New York and Massachusetts (Cramp & Perrins 1994b).

From May to July 2009, no less than three birds occurred in Britain (Hudson & Rarities Committee 2010; Millington 2009; Shaw 2009), followed by the first for France and another in Britain in May 2010 (van den Berg & Haas 2010c).

2 records

Britain (1)
1 • 24 April 1988, Ardnave, Islay, Strathclyde, male [British Ornithologists' Union 1993b; McKay 1994; Rogers & Rarities Committee 1993]

Norway (1)
1 • 01 June 1987, Jomfruland, Kragerø, Telemark, adult female, trapped, died (ZMO: 11452) [Bentz 1989; Gustad 1995b; Nicolaisen 1987; Ree 2009]

1 June 1987, Jomfruland, Kragerø, Telemark, Norway. © Jan T Lifjeld

Yellow-headed Blackbird *Xanthocephalus xanthocephalus*

Breeds in North America, from central British Columbia and northern Alberta east through central Manitoba to north-western Ohio, south to southern California, southern New Mexico, north-eastern Mexico, south-western Missouri and central Illinois. Winters from southern North America to southern Mexico (Cramp & Perrins 1994b). It was also recorded in Greenland in September 1840 and August 1900 (Boertmann 1994).

Proctor & Donald (2003) mention more records (for instance, several from Britain), but nearly all have been rejected or placed in category D (cf. British Ornithologists' Union 1994a, 2006a). The two records given below are the only ones on their list that have been accepted as genuine vagrants.

2 records

Iceland (1)

1 • 23–24 July 1983, Hafnarnes í Nesjum, Austur-Skaftafellssýsla, collected (IMNH: RM8120) [Ólafsson 1993b; Petersen 1985; Pétursson & Ólafsson 1985]

Netherlands (1)

1 • 18 and 20 May and 14 June 1982, Polder Waal en Burg, Texel, Noord-Holland, adult [Ebels & van den Bergh 2007; van Vliet & Ebels 2007; van der Vliet *et al.* 2007]
• 02–03 July 1982, Formerumer Wiel, Terschelling, Friesland, adult

24 July 1983, Hafnarnes í Nesjum, Austur-Skaftafellssýsla, Iceland. © Yann Kolbeinsson

Golden-winged Warbler *Vermivora chrysoptera*

Breeds in eastern North America, from Minnesota east through Great Lakes region to New England and south to northern Georgia. Winters mainly in Central America, from Guatemala to Panama (Curson *et al.* 1994). It was also recorded in Greenland in the autumn of 1966 (Boertmann 1994).

This sole record is of a bird that spent the winter at an inland location. Doherty (1992) suggests that it probably arrived the previous autumn.

1 record

Britain (1)

1 • 24 January to 10 April 1989, Larkfield, Maidstone, Kent, male [British Ornithologists' Union 1991a; Doherty 1992; Rogers & Rarities Committee 1990]

February 1989, Larkfield, Maidstone, Kent, Britain. © David Cottridge

Blue-winged Warbler *Vermivora pinus*

Breeds in eastern North America, from Minnesota east to New England and northern Georgia. Winters mainly in Central America, from southern Mexico south to Costa Rica (Curson *et al.* 1994).

The sole record is from Cape Clear Island, a hotspot for Nearctic vagrants where various other vagrant species have been recorded. The timing of the record fits in well with the migration pattern of this species.

1 record

Ireland (1)
1 • 04–10 October 2000, Cape Clear Island, Cork, first-year male [Milne *et al.* 2002; Wing 2000]

October 2000, Cape Clear Island, Cork, Ireland. © Phil Palmer

Tennessee Warbler *Vermivora peregrina*

Breeds across northern North America, from southern Yukon, British Columbia and southern Alaska and north-western Montana, east to Newfoundland and northern New England. Winters from southern Mexico south to Colombia and northern Ecuador and Venezuela (Curson *et al.* 1994). It was also recorded in Greenland in August 1840, 1898 and in the autumn of 1991 (Boertmann 1994).

A record in France in April 1997 was rejected after review (Frémont & CHN 1999; Reeber *et al.* 2008).

September 1984, Sumba, Suðuroy, Faeroes.
© Søren Sørensen

7 September 1982, Holm, Orkney, Britain.
© Eric Meek

7 records

Azores (1)
1 • 21 November 2005, Fojo, Corvo, first-winter [Jara *et al.* 2008]

Britain (4)
1 • 06–20 September 1975, Fair Isle, Shetland, immature, trapped (18 September) [British Ornithologists' Union 1978; Broad 1981; Dymond & Rarities Committee 1976; Rogers & Rarities Committee 1981]
2 • 24 September 1975, Fair Isle, Shetland, immature, trapped [British Ornithologists' Union 1978; Broad 1981; Dymond & Rarities Committee 1976]
3 • 05–07 September 1982, Holm, Orkney, first-year, trapped (07 September) [Meek 1984; Rogers & Rarities Committee 1983]
4 • 20 September 1995, Hirta, St Kilda, Outer Hebrides [Rogers & Rarities Committee 1996]

Faeroes (1)
1 • 21–28 September 1984, Sumba, Suðuroy, first-year, (re)trapped (21 September and on two other days) [Boertmann *et al.* 1986]

Iceland (1)
1 • 14 October 1956, Hallbjarnareyri í Eyrarsveit, Snæfellsnessýsla, immature, found dead (IMNH: RM5578) [Ólafsson & Pétursson 1997; Pétursson & Þráinsson 1999]

18 September 1975, Fair Isle, Shetland, Britain. © Tony Broome

Chestnut-sided Warbler *Dendroica pensylvanica*

Breeds in central-eastern North America, from Saskatchewan east to Nova Scotia and south through Great Lakes region and New England, and also to northern Georgia and Alabama. Winters mainly in southern Central America, from Nicaragua to Panama (Curson *et al.* 1994). It was also recorded in Greenland in the winter of 1887, the summer of 1955 and September 1974 (Boertmann 1994).

Both records occurred in record years for Nearctic vagrants; the first Azores bird was found in October 2009 (Gantlett 2010).

2 records

Britain (2)
1. • 20 September 1985, Fetlar, Shetland, first-year [British Ornithologists' Union 1991a; Peacock 1993; Rogers & Rarities Committee 1988]
2. • 18 October 1995, Prawle Point, Devon [Brett 1995; Rogers & Rarities Committee 1996]

Cerulean Warbler *Dendroica cerulea*

Breeds in eastern North America, from Great Lakes region east to New England and south to Louisiana and Georgia. Winters mainly in western South America, from Columbia and Venezuela south to Peru and Bolivia (Curson *et al.* 1994).

This record occurred at the same time as three other new Parulidae were recorded in south-western Iceland (cf. Hjartarson *et al.* 2003) in September–October 1997.

1 record

Iceland (1)

1 • 01–07 October 1997, Eyrarbakki, Árnessýsla, immature female, trapped (07 October) [Hjartarson *et al.* 2003; Kolbeinsson *et al.* 2000; Þráinsson 1997]

October 1997, Eyrarbakki, Árnessýsla, Iceland. © Jóhann Óli Hilmarsson

Black-throated Blue Warbler *Dendroica caerulescens*

Breeds in central eastern North America, from western Ontario and Minnesota east to Nova Scotia and New England; also south in Appalachians to northern Georgia. Winters in West Indies, mainly Bahamas and Greater Antilles; a few winter in Florida (Curson *et al.* 1994). It was also recorded in Greenland in September 1964, September 1965 and October 1988 (Boertmann 1994).

A record from the Azores in October–November 2005 is still under review by the rarities committee (João Jara in litt).

4 records

Azores (2)

1 • 24–29 October 2006, Corvo, first-winter female [João Jara in litt]
2 • 28 October 2006, Corvo, first-winter male [João Jara in litt]

Iceland (2)

1 • 14–19 September 1988, Heimaey, Vestmannaeyjar, adult male, collected (IMNH: RM9725) [Ólafsson & Pétursson 1997; Petersen 1989; Pétursson *et al.* 1991]
2 • 17–18 October 2003, Heimaey, Vestmannaeyjar, adult female [Kolbeinsson *et al.* 2006]

24 October 2006, Corvo, Azores. © Vincent Legrand

Extremely rare birds in the Western Palearctic

Black-throated Blue Warbler *Dendroica caerulescens*

28 October 2006, Corvo, Azores. © Vincent Legrand

19 September 1988, Heimaey, Vestmannaeyjar, Iceland. © Yann Kolbeinsson

October 2003, Heimaey, Vestmannaeyjar, Iceland. © Daniel Bergmann

Black-throated Green Warbler *Dendroica virens*

Breeds in North America, from extreme eastern British Columbia east to Newfoundland and New England, and south in Appalachians to central Georgia and Alabama; also in coastal South and North Carolina. Winters from southern Florida and south-eastern Texas, south through Central America to Panama; also in West Indies (Curson *et al.* 1994). It was also recorded in Greenland in 1853, the autumn of 1933 and September 1949 (Boertmann 1994).

A first-winter female found dead on board a ship at Sundahöfn, Reykjavík, in September 1984 has not been accepted onto the Icelandic list (cf. Ólafsson & Pétursson 1997). In October 2007, one landed on a ship 190 nautical miles south-west of the Azores (outside Western Palearctic waters) and died the next day (Gantlett 2008). The first to third records for the Azores occurred in October 2008 and October 2009 (two) (Gantlett 2009, 2010).

19 October 1858, Helgoland, Schleswig-Holstein, Germany. © Jochen Dierschke

27 October 2003, Þorbjörn við Grindavík, Gullbringusýsla, Iceland. © Yann Kolbeinsson

2 records

Germany (1)

1 • 19 October 1858, Helgoland, Schleswig-Holstein, first-winter male, collected (IVVH: C4182) [Dierschke 1998; Gätke 1900]

Iceland (1)

1 • 27–28 October 2003, Þorbjörn við Grindavík, Gullbringusýsla, immature male [Kolbeinsson 2006; Kolbeinsson *et al.* 2006]

Blackburnian Warbler *Dendroica fusca*

Breeds in eastern North America, from Saskatchewan east to Newfoundland and New England and south in Appalachians to western North Carolina and eastern Tennessee. Winters in Central and South America, from Colombia, Venezuela and Brazil south to Bolivia (Curson *et al.* 1994). It was also recorded in Greenland in September 1953 and the autumn of 1985 (Boertmann 1994).

Both British records occurred in the first week of October. The first Western Palearctic bird from October 1961 was identified after having been published as an unidentified *Dendroica* warbler (cf. Nisbet 1963b). The third for Britain occurred in September 2009 (Hudson & Rarities Committee 2010; Millington 2009).

3 records

Britain (2)

1 • 05 October 1961, Skomer, Pembrokeshire [British Ornithologists' Union 1991a; Rogers & Rarities Committee 1990; Saunders & Saunders 1992]

2 • 07 October 1988, Fair Isle, Shetland, first-winter male [British Ornithologists' Union 1991a; Rogers & Rarities Committee 1990; Willmott 1988]

Iceland (1)

1 • between 06 and 16 September 1987 or between 09 and 19 November 1987, Strandagrunn, c 40 nautical miles north-east off Horni, c 66° 50′ N 21° 25′ W, at sea, first-winter female, found moribund (on board a ship) and collected (IMNH: RM9949) [Ólafsson & Pétursson 1997; Petersen 1989; Pétursson & Ólafsson 1989]

Between 6-16 September or 9-19 November 1987, off Horni, at sea, Iceland. © Yann Kolbeinsson

SPECIES ACCOUNTS

Cape May Warbler *Dendroica tigrina*

Breeds in northern North America, from Northwest Territories and British Columbia east to Nova Scotia and New England, including northern Great Lakes region. Winters mainly in West Indies (Curson *et al*. 1994).

A remarkable record of a singing male in June.

1 record

Britain (1)
1 • 17 June 1977, Paisley, Clyde, male [British Ornithologists' Union 1980; Byars & Galbraith 1980; Rogers & Rarities Committee 1979]

17 June 1977, Paisley, Clyde, Britain. © Tony Byar

Magnolia Warbler *Dendroica magnolia*

Breeds in northern North America, from south-western Northwest Territories and British Columbia east to Newfoundland and New England, and south in Appalachians to western Virginia. Winters in Central America, from southern Mexico to Nicaragua; also in small numbers in West Indies (Curson *et al.* 1994). It was recorded in Greenland in the autumn of 1875, May 1880 and October 1950 (Boertmann 1994).

In the Azores, a report in September 1999 (Gantlett 2000) has not (yet) been submitted to the Portuguese rarities committee (João Jara in litt). A bird photographed in October 2009 constituted the first definite record for the Azores (Gantlett 2010).

3 records

Britain (1)
1 • 27–28 September 1981, St Agnes, Isles of Scilly [British Ornithologists' Union 1984; Enright 1995; Rogers & Rarities Committee 1982]

Iceland (2)
1 • 29 September to 07 December 1995, Bakki í A-Landeyjum, Rangárvallasýsla, immature, trapped (08 November) [Ólafsson & Pétursson 1997; Þráinsson 1998; Þráinsson & Pétursson 1997]
2 • 21–23 October 1995, Seltjörn við Njarðvík, Gullbringusýsla, immature [Ólafsson & Pétursson 1997; Þráinsson 1998; Þráinsson & Pétursson 1997]

26 November 1995, Bakki í A-Landeyjum, Rangárvallasýsla, Iceland. © Örn Óskarsson

SPECIES ACCOUNTS

October 1995, Seltjörn við Njarðvík, Gullbringusýsla, Iceland. © Yann Kolbeinsson

Palm Warbler *Dendroica palmarum*

Breeds in northern North America, from Northwest Territories east to Newfoundland and New England. Winters mainly in West Indies and on Atlantic coast (Curson *et al.* 1994).

This bird was assigned to the eastern subspecies *hypochrysea*, which suggests a genuine occurrence. It was also recorded at a time when several other Parulidae were recorded in south-western Iceland (cf. Hjartarson *et al.* 2003). A record of a headless corpse on Walney Island, Cumbria, Britain, in May 1976 was not accepted onto the British list because tideline corpses are not accepted into category A (cf. British Ornithologists' Union 1978, 2001; O'Sullivan & Rarities Committee 1977).

1 record

Iceland (1)

1 • 5–10 October 1997, Stokkseyri, Árnessýsla [Hjartarson *et al.* 2003; Kolbeinsson *et al.* 2000; Þráinsson 1997]

7 October 1997, Stokkseyri, Árnessýsla, Iceland. © Örn Óskarsson

Bay-breasted Warbler *Dendroica castanea*

Breeds in northern North America, from Northwest Territories and British Columbia east to Newfoundland, Nova Scotia and New England, including northern Great Lakes region. Winters mainly in South America, from Panama to Colombia and Venezuela north to Costa Rica (Curson *et al.* 1994). It wao recorded in Greenland in October 1898 (Boertmann 1994).

This bird was videoed at a time when several other Nearctic species were recorded in western Europe and was identified after study of the footage.

1 record

Britain (1)

1 • 01 October 1995, Land's End, Cornwall, first-winter male [British Ornithologists' Union 1998; Ferguson 1997; Rogers & Rarities Committee 1997]

Ovenbird *Seiurus aurocapilla*

Breeds in North America, from eastern British Columbia and Alberta east to Newfoundland, and south to Colorado in the west and to Oklahoma and South Carolina in the east. Winters from extreme southern Northern America (Gulf coast and Florida) south through Central America to Panama, in West Indies and south to northern Colombia and Venezuela (Curson *et al.* 1994). It was recorded in Greenland in October 1933, 1943, March 1946 and October 1954 (Boertmann 1994).

A wing found at Formby, Lancashire, in January 1969, has not been accepted onto the British list (cf. Smith & Rarities Committee 1970; British Ornithologists' Union 1971b). Up to two occurred in the Azores in October–November 2009 (Gantlett 2010).

21 October 2008, Serreta, Terceira, Azores. © Ferran López

Ovenbird *Seiurus aurocapilla*

9 records (10 individuals)

Azores (2/3)
1 • 01 November to at least 25 December 2005, Corvo [Jara *et al.* 2007]
2 • 21 October to 18 November 2008, Serreta, Terceira, *maximum* of two [João Jara in litt]

Britain (4)
1 • 07–08 October 1973, Out Skerries, Shetland, trapped (07 October) [British Ornithologists' Union 1974; Robertson 1975; Smith & Rarities Committee 1974]
2 • 22 October 1985, Spriddlestone, Wembury, Devon, probably first-winter, found dead (private collection) [Rogers & Rarities Committee 1986; Ward 1987]
3 • 20 December 2001 to 16 February 2002, Much Marcle, Herefordshire [Rogers & Rarities Committee 2002]
4 • 25–28 October 2004, St Mary's, Isles of Scilly, taken into care, died on 28 October (ISM: RN 05937) [Harding 2004; Rogers & Rarities Committee 2005]

Ireland (2)
1 • 08 December 1977, Lough Carra Forest, Mayo, found dead (specimen location unknown) [Preston 1981; Wilson 1980]
2 • 24–25 September 1990, Dursey Island, Cork, first-winter [Grace & Lancaster 1990; O'Sullivan & Smiddy 1991]

Norway (1)
1 • 25–27 October 2003, Beiningen, Karmøy, Rogaland [Mjølsnes *et al.* 2005]

October 1973, Out Skerries, Shetland, Britain.
© Ian Robertson

SPECIES ACCOUNTS

December 2001, near Much Marcle, Herefordshire, Britain. © Anonymous

September 1990, Dursey Island, Cork, Ireland. © Anthony McGeehan

26 October 2003, Beiningen, Karmøy, Rogaland, Norway. © Morten Vang

Louisiana Waterthrush *Seiurus motacilla*

Breeds in south-eastern North America, excluding Florida, from Great Lakes region and New England to Kansas and Texas. Winters mainly in Central and South America, from Mexico south to Panama and in West Indies (Curson *et al.* 1994). It was recorded in Greenland in 1949 (Boertmann 1994).

After a published photograph (Birding World 5: 26, 1992) was labelled as Northern Waterthrush *S. noveboracensis*, the true identity of this bird was established and corrected nearly eight years later in Gutiérrez & Lorenzo (1999).

1 record

Canary Islands (1)

1 • 10–26 November 1991, Tazacorte, Playa Nueva, La Palma [Gutiérrez & Lorenzo 1999; de Juana 2006; de Juana & Comité de Rarezas de la SEO 1998]

November 1991, Tazacorte, Playa Nueva, La Palma, Canary Islands. © Martin Semisch

Hooded Warbler *Wilsonia citrina*

Breeds in south-eastern North America, west to Iowa and eastern Texas, and north to southern Great Lakes region and southern New England. Winters in Central America, from south-eastern Mexico to Costa Rica, rarely to Panama; a few winter in north-western Mexico and in West Indies (Curson *et al.* 1994).

The first British record is not recognized by Evans (1994), although it is still accepted by the British rarities committee.

4 records

Azores (2)
1 • 26–27 October 2005, Ribeira do Gavião, Corvo, male [Jara *et al.* 2007]
2 • 11 October 2008, Corvo, male [João Jara in litt]

Britain (2)
1 • 20–23 September 1970, St Agnes, Isles of Scilly, immature or female [British Ornithologists' Union 1972; Edwards & Osborne 1972; Smith & Rarities Committee 1972]
2 • 10 September 1992, Hirta, St Kilda, Outer Hebrides, probably immature male [Dix 1992; Rogers & Rarities Committee 1993]

26 October 2005, Ribeira do Gavião, Corvo, Britain. © Peter Alfrey

11 October 2008, Corvo, Azores. © Dominic Mitchell/ www.birdingetc.com

Wilson's Warbler *Wilsonia pusilla*

Breeds in northern North America, from Alaska east to Newfoundland and Nova Scotia and south to California and New Mexico. Winters in Central America, from Mexico south to Panama (Curson *et al.* 1994). It was also recorded in Greenland in September 1975 (Boertmann 1994).

The sole record occurred in a record autumn for Nearctic vagrants (cf. Dawson & Allsopp 1986).

1 record

Britain (1)

1 • 13 October 1985, Rame Head, Cornwall, male [British Ornithologists' Union 1991a; Rogers & Rarities Committee 1988, 1989; Smaldon 1990]

Canada Warbler *Wilsonia canadensis*

Breeds in northern North America, from British Columbia and Alberta east to Nova Scotia, including Great Lakes region. Winters mainly in South America, from Colombia and Venezuela south to Peru (Curson *et al.* 1994). It was recorded in Greenland in 1875, October 1943 and September 1981 (Boertmann 1994).

The first for the Azores occurred in October 2009 (Gantlett 2010).

2 records

Iceland (1)

1 • 29 September 1973, Sandgerði, Gullbringusýsla, male, found moribund and collected (IMNH: RM5589) [Ólafsson & Pétursson 1997; Pétursson & Þráinsson 1999]

Ireland (1)

1 • 08–13 October 2006, Kilbaha, Loop Head, Clare, first-winter female [Hanafin 2006; Milne & McAdams 2008]

SPECIES ACCOUNTS

10 October 2006, Kilbaha, near Loop Head, Clare, Ireland. © Chris Batty

29 September 1973, Sandgerði, Gullbringusýsla, Iceland. © Yann Kolbeinsson

209

REFERENCES

Addinall, S. 2005. The Amur Wagtail in County Durham - a new Western Palearctic bird. *Birding World* 18: 155-158.

Addinall, S. G. 2010. 'Amur Wagtail' in County Durham: new to Britain and the Western Palearctic. *British Birds* 103: 260-267.

Ahmad, M. 2007. The Pacific Diver in Cornwall – the third for the Western Palearctic. *Birding World* 20: 62-63.

Al Hajji, R., Alsirhan, A., Fagel, P., Foster, B. & Pope, M. 2008. The Purple Sunbirds in Kuwait – a new Western Palearctic bird. *Birding World* 21: 71-74.

Aley, J. & Aley, R. 1995. Red-breasted Nuthatch in Norfolk: new to Britain and Ireland. *British Birds* 88: 150-153.

Aley, P. H. 1987. Veery on Lundy - The second British Record. *Twitching* 1: 371-373.

Allan, J. 2007. The Glaucous-winged Gull in Surrey. *Birding World* 20: 150-151.

Allsopp, E. M. P. 1972. Veery in Cornwall: a species new to Britain and Ireland. *British Birds* 65: 45-49.

Alström, P. & Mild, K. 2003. *Pipits and wagtails of Europe, Asia and North America*. London.

Andersen, B. A. 1974. Nordamerikansk fugleart i Onsøy. *Østfoldornitologen* 1: 25-26.

Andersson, G. 1971. Dvargspov, *Numenius minutus*, i Finnmark. *Sterna* 10: 63-64.

Andrews, I. J. 1994. Description and status of the black morph Mourning Wheatear *Oenanthe lugens* in Jordan. *Sandgrouse* 16: 32-35.

Andrews, I. J., Khoury, F. & Shirihai, H. 1999. Jordan Bird Report 1995-97. *Sandgrouse* 21: 10-35.

Anonymous. 1954. Magnificent Frigate-bird in Tiree, Inner Hebrides: a new British bird. *British Birds* 47: 58-59.

Anonymous. 1992. From the Archives: the Ruby-crowned Kinglet on Loch Lomondside. *Birding World* 5: 195-199.

Anonymous. 1998a. Bird News October 1998. *Birding World* 11: 366-377.

Anonymous. 1998b. The Mourning Dove at Heathrow – a plane-assisted vagrant. *Rare Birds* 4: 105.

Anonymous. 2003. Western Palearctic news. *Birding World* 16: 276-280.

Anonymous. 2005. Bird News October 2005. *Birding World* 18: 398-423.

Antonucci, A. & Corso, A. 2008. The Willet in Italy – a new bird for the Mediterranean Basin. *Birding World* 21: 75-79.

Arkhipov, V. Y., Haas, M. & Crochet, P.-A. 2010. Western Palearctic list updates: Yellow-eyed Dove. *Dutch Birding* 32: 191-193.

Aspinall, S. 1987. Savannah Sparrow on Fair Isle - The Second British Record. *Twitching* 1: 337-338.

Astins, D. & Brown, J. 2007. The Pacific Diver in Pembrokeshire – the second for the Western Palearctic. *Birding World* 20: 57-61.

Aumüller, R. 2005. Der Gelbkehlvireo *Vireo flavifrons*, eine neue Art für Deutschland. *Limicola* 19: 323-329.

Baha El Din, S. M. 1992. Ornithological news from Egypt. *The Courser* 3: 59-67.

Baines, R. 2007. The Brown Flycatcher in East Yorkshire. *Birding World* 20: 425-428.

Baker, N. E. 1989. Shy Albatross *Diomedea cauta*: the first record for Tanzania. *Scopus* 13: 115.

Bakker, T., van Dijken, K. & Ebels, E. E. 2001. Glaucous-winged Gull at Essaouira, Morocco, in January 1995. *Dutch Birding* 23: 271-274.

Balogh, G. 1997. Spectacles on ice. *Birding World* 10: 103-107.

Bannerman, D. A. & Bannerman, W. M. 1965. *Birds of the Atlantic Islands*. Volume 2. A history of the birds of Madeira, the Desertas, and the Porto Santo Islands. Edinburgh.

Bannerman, D. A. & Bannerman, W. M. 1966. *Birds of the Atlantic Islands*. Volume 3. A history of the birds of the Azores. London.

Barthel, P. H. & Helbig, A. J. 2005. Artenliste der Vögel Deutschlands. *Limicola* 19: 89-111.

Bashford, R. 1997. The first Cotton Teal *Nettapus coromandelianus* in Jordan. *Sandgrouse* 19: 142-143.

Beaman, M. 1990. Identification of Black Noddy and validity of December 1984 record in Mauritania. *Dutch Birding* 12: 245-248.

Bear, A. 1991. Ethiopian Swallow *Hirundo aethiopica*: A new species for Israel and the Palearctic region. *Torgos* 9 (2): 73.

Bentz, P.-G. 1987. Sjeldne fugler i Norge i 1985. *Vår Fuglefauna* 10: 91-95.

Bentz, P.-G. 1988. Sjeldne fugler i Norge i 1986. *Vår Fuglefauna* 11: 87-95.

Bentz, P.-G. 1989. Sjeldne fugler i Norge i 1987. *Vår Fuglefauna* 12: 101-110.

van den Berg, A. B. 2000. WP reports - March-April 2000. *Dutch Birding* 22: 112-121.

van den Berg, A. B. 2004a. Population growth and vagrancy potential of Ross's Goose. *Dutch Birding* 26: 107-111.

van den Berg, A. B. 2004b. Spectacled Eider's window of time. *Dutch Birding* 26: 314-318.

van den Berg, A. B. & Bosman, C. A. W. 2001. *Zeldzame vogels van Nederland – Rare birds in the Netherlands*. Avifauna van Nederland 1. Second edition. Haarlem.

van den Berg, A. B., de By, R. A. & CDNA 1989. Rare birds in the Netherlands in 1988. *Dutch Birding* 11: 151-164.

van den Berg, A. B., de By, R. A. & CDNA 1991. Rare birds in the Netherlands in 1989. *Dutch Birding* 13: 41-57.

van den Berg, A. B., de By, R. A. & CDNA 1992. Rare birds in the Netherlands in 1990. *Dutch Birding* 14: 73-90.

van den Berg, A. B., de By, R. A. & CDNA 1993. Rare birds in the Netherlands in 1991. *Dutch Birding* 15: 145-159.

van den Berg, A. B. & Cottaar, F. 1986. Ross Gans in Noordholland in november-december 1985. *Dutch Birding* 8: 57-59.

REFERENCES

van den Berg, A. B., Douma, F. & Kuiken, D. 1993. Canadese Kraanvogel te Paesens-Moddergat in september 1991. *Dutch Birding* 15: 1-6.

van den Berg, A. B. & Haas, M. 2006. WP reports – late September-early November 2006. *Dutch Birding* 28: 371-390.

van den Berg, A. B. & Haas, M. 2007. WP reports – late September-November 2007. *Dutch Birding* 29: 378-400.

van den Berg, A. B. & Haas, M. 2009. WP reports – late March-mid-May 2009. *Dutch Birding* 31: 186-198.

van den Berg, A. B. & Haas, M. 2010a. WP reports – late November 2009 to mid-January 2010. *Dutch Birding* 32: 52-62.

van den Berg, A. B. & Haas, M. 2010b. WP reports – late January to mid-March 2010. *Dutch Birding* 32: 134-142.

van den Berg, A. B. & Haas, M. 2010c. WP reports – late March to late May 2010. *Dutch Birding* 32: 199-214.

van den Berg, A. B. & Haas, M. 2010d. WP reports – June to late July 2010. *Dutch Birding* 32: 266-276.

Bergier, P., Franchimont, J., Thévenot, M. & CHM 1997. Les oiseaux rares au Maroc. Rapport de la Commission d'Homologation Marocaine. Numéro 2. *Porphyrio* 9: 165-173.

Bergier, P., Franchimont, J., Thévenot, M. & CHM 2000. Rare birds in Morocco: report of the Moroccan Rare Birds Committee (1995-1997). *African Bird Club Bull* 7: 18-28.

Bergier, P., Franchimont, J., Thévenot, M. & CHM 2008. Les oiseaux rares au Maroc. Rapport de la Commission d'Homologation Marocaine, Numéro 13. *Go-South Bulletin* 5: 48-58.

Bergier, P., Franchimont, J., Thévenot, M. & CHM 2010. Les oiseaux rares au Maroc. Rapport de la Commission d'Homologation Marocaine, Numéro 15 (2009). *Go-South Bulletin* 7: 1-14.

Bergier, P., Zadane, Y. & Qninba, A. 2009. Cape Gull: a new breeding species in the Western Palearctic. *Birding World* 22: 253-256.

Berlijn, M. 2004. Ross' Ganzen in Nederland in 1988-2003. *Dutch Birding* 26: 100-106.

Bezzel, E. 1987. Keilschwanzsturmtaucher *(Puffinus pacificus)* in Ägypten: "Neue" Art für die Westpläarktis? *Die Vogelwelt* 108: 71-72.

Bigas, D. & Gutiérrez, R. 2000. The Siberian Blue Robin in Spain - the second Western Palearctic record. *Birding World* 13: 415-417.

Birch, A. 1990. Yellow-throated Vireo in Cornwall - A new Western Palearctic bird. *Birding World* 3: 308-309.

Birch, A. 1994. Yellow-throated Vireo: new to Britain and Ireland. *British Birds* 87: 362-365.

Blake-Knox, H. 1870. Esquimaux Curlew in Dublin Market. *The Zoologist* 5 (second series): 2408-2409.

Blankert, J. J., de By, R. A. & CDNA 1987. Rare birds in the Netherlands in 1986. *Dutch Birding* 9: 143-151.

Blankert, J. J., Scharringa, C. J. G. & CDNA 1984. Rare birds in the Netherlands in 1982. *Dutch Birding* 6: 45-53.

Blomdahl, A. 2006. Sällsynta fåglar i Sverige 2005 - rapport från SOF:s raritetskommitté. *Vår Fågelvärld Supplement* 45: 63-159.

Blomdahl, A. & Stridh, T. 2007. Sällsynta fåglar i Sverige 2006 - rapport från SOF:s raritetskommitté. *Vår Fågelvärld Supplement* 47: 57-151.

BMAPT & KEPS 2007. Kuwait Bird Report 2006. Website: http://www.hawar-islands.com/kuwait_birding/BMAPT_2006.html.

Boano, G. & Mingozzi, T. 1986. Gli uccelli di comparsa accidentale nella regione piemontese. Nota complementare. *Rivista Piemontese di Storia Naturale* 7: 217-218.

Boertmann, D. 1994. *An annotated checklist to the birds of Greenland*. Meddelelser om Grønland. Bioscience 38. Copenhagen.

Boertmann, D., Sørensen, S. & Pihl, S. 1986. Sjældne fugle på Færøerne i årene 1982-1985. *Dansk Orn Foren Tidsskr* 80: 121-130.

Boesman, P. 1990. Indian Pond Herons. *Birding World* 3: 251.

Bolt, D. 2001. The Redhead in Glamorgan. *Birding World* 14: 495-496.

Bonser, R. 2008. The Tricolored Heron on the Canary Islands. *Birding World* 21: 67-70.

Borg, J. J. (in press). *The Catalogue to the Ornithological Collections of the National Museum of Natural History*, Mdina. Mdina.

Bot, M., Ebels, E. B. & Pohlmann, H. 2010. Taigastrandloper bij Zwolle in oktober 2009. *Dutch Birding* 32: 316-320.

Bourne, W. R. P. 1966. Observations of sea-birds. *Sea Swallow* 18: 9-36.

Bourne, W. R. P. 1967. Long-distance vagrancy in petrels. *Ibis* 109: 141-167.

Bourne, W. R. P. 1989. McCormick's Skua in North-West European waters. *Sea Swallow* 38: 63-64.

Bourne, W. R. P. 1992. Atlantic Petrel in the Western Palearctic. *Dutch Birding* 14: 100-101.

Bouwman, R. G. 1985. Spur-winged Goose in Morocco in June 1984. *Dutch Birding* 7: 21-22.

Brazier, H., Dowdall, J. F., Fitzharris, J. E. & Grace, K. 1986. Thirty-third Irish Bird Report, 1985. *Irish Birds* 3: 287-336.

Breen, D. 2008. The Little Blue Heron in County Galway – a new European bird. *Birding World* 21: 436-440.

Breife, B., Hirschfeld, E., Kjellén, N. & Ullman, M. 2003. *Sällsynta fåglar i Sverige*. Stockholm.

Brett, A. 1995. The Chestnut-sided Warbler in Devon. *Birding World* 8: 391.

Brichetti, P., Arcamone, E. & COI 1995. Comitato di Omologazione Italiano (C. O. I.) 10. *Rivista Italiana di Ornitologia* 65: 147-149.

Brichetti, P., Arcamone, E. & COI 1996. Comitato di Omologazione Italiano (C. O. I.) 11. *Rivista Italiana di Ornitologia* 66: 171-174

Brichetti, P., Arcamone, E. & COI 1997. Comitato di Omologazione Italiano (C. O. I.) 12. *Rivista Italiana di Ornitologia* 67: 189-192.

Brichetti, P., Arcamone, E., Occhiato, D. & COI 2002. Commissione Ornitologica Italiana (COI) già Comitato di Omologazione Italiano. Report 15. *Avocetta* 26: 117-121.

Bried, J. 2003. A Bermuda Petrel on the Azores - a new Western Palearctic bird. *Birding World* 16: 22.

British Ornithologists' Union 1956. British Records Sub-Committee: First Report. *Ibis* 98: 154-157.

British Ornithologists' Union 1958. British Records Committee: Second Report (December 1957). *Ibis* 100: 299-300.

British Ornithologists' Union 1963. British Records Committee: Fourth Report (February 1963). *Ibis* 105: 289-291.

British Ornithologists' Union 1971a. British Records Committee: Fifth Report (December 1968). *Ibis* 113: 142-145.

British Ornithologists' Union 1971b. Records Committee: Sixth Report (February 1971). *Ibis* 113: 420-423.

British Ornithologists' Union 1972. Records Committee: Seventh Report (April 1972). *Ibis* 114: 446-447.

British Ornithologists' Union 1974. Records Committee: Eighth Report (March 1974). *Ibis* 116: 578-579.

British Ornithologists' Union 1978. Records Committee: Ninth Report (April 1978). *Ibis* 120: 409-411.

British Ornithologists' Union 1980. Records Committee: Tenth Report (March 1980). *Ibis* 122: 564-568.

British Ornithologists' Union 1984. Records Committee: Eleventh Report (December 1983). *Ibis* 126: 440-444.

British Ornithologists' Union 1986. Records Committee: Twelfth Report (April 1986). *Ibis* 128: 601-603.

British Ornithologists' Union 1988. Records Committee: Thirteenth Report (December 1987). *Ibis* 130: 334-337.

British Ornithologists' Union 1991a. Records Committee: Fourteenth Report (August 1990). *Ibis* 133: 218-222.

British Ornithologists' Union 1991b. Records Committee: Fifteenth Report (April 1991). *Ibis* 133: 438-441.

British Ornithologists' Union 1992a. Records Committee: Sixteenth Report (December 1991). *Ibis* 134: 211-214.

British Ornithologists' Union 1992b. Records Committee: Seventeenth Report (May 1992). *Ibis* 134: 380-381.

British Ornithologists' Union 1993a. Records Committee: Eighteenth Report (December 1992). *Ibis* 135: 220-222.

British Ornithologists' Union 1993b. Records Committee: Nineteenth Report (May 1993). *Ibis* 135: 493-499.

British Ornithologists' Union 1994a. Records Committee: Twentieth Report (December 1993). *Ibis* 136: 253-255.

British Ornithologists' Union 1994b. Records Committee: Twenty-first Report (May 1994). *Ibis* 136: 497.

British Ornithologists' Union 1995. Records Committee: Twenty-second Report (May 1995). *Ibis* 137: 590-591.

British Ornithologists' Union 1997. Records Committee: Twenty-third Report (July 1996). *Ibis* 139: 197-201.

British Ornithologists' Union 1998. Records Committee: Twenty-fourth Report (October 1997). *Ibis* 140: 182-184.

British Ornithologists' Union 1999. Records Committee: 25th Report (October 1998). *Ibis* 141: 175-180.

British Ornithologists' Union 2001. Records Committee: 27th Report (October 2000). *Ibis* 143: 171-175.

British Ornithologists' Union 2002. Records Committee: 28th Report (October 2001). *Ibis* 144: 181-184.

British Ornithologists' Union 2003. Records Committee: 29th Report (October 2002). *Ibis* 145: 178-183.

British Ornithologists' Union 2004. Records Committee: 30th Report (October 2003). *Ibis* 146: 192-195.

British Ornithologists' Union 2005. Records Committee: 31st Report (October 2004). *Ibis* 147: 246-250.

British Ornithologists' Union 2006a. Records Committee: 32nd Report (October 2005). *Ibis* 148: 198-201.

British Ornithologists' Union 2006b. Records Committee: 33rd Report (April 2006). *Ibis* 148: 594.

British Ornithologists' Union 2007a. Records Committee: 34th Report (October 2006). *Ibis* 149: 194-197.

British Ornithologists' Union 2007b. Records Committee: 35th Report (April 2007). *Ibis* 149: 652-654.

British Ornithologists' Union 2009. Records Committee: 37th Report (October 2008). *Ibis* 151: 224-230.

British Ornithologists' Union 2010. Records Committee: 38th Report (October 2009). *Ibis* 152: 199-204.

British Ornithologists' Union 2011. Records Committee: 39th Report (October 2010). *Ibis* 153: 227-232.

Broad, R. A. 1981. Tennessee Warblers: new to Britain and Ireland. *British Birds* 74: 90-94.

Brodie Good, J. 1991. Philadelphia Vireo in Scilly: new to Britain. *British Birds* 84: 572-574.

Brodie Good, J. & Filby, R. A. 1987. A new British bird - Philadelphia Vireo on Tresco. *Twitching* 1: 301-302.

Brouwer, R. E. & Halff, R. A. C. 1999. Daurische Kauw in Noord-Holland in mei 1997. *Dutch Birding* 21: 151-153.

Brown, P. 2001. The Siberian Blue Robin on Orkney. *Birding World* 14: 422-423.

Broyd, S. J. 1985. Savannah Sparrow: new to the Western Palearctic. *British Birds* 78: 647-656.

Bunes, V. & Solbakken, K. A. 2004. Sjeldne fugler i Norge i 2002. *Ornis Norvegica* 27: 4-47.

de By, R. A., van den Berg, A. B. & CDNA 1992. Zeldzame en schaarse vogels in Nederland in 1990. *Limosa* 65: 137-146.

de By, R. A., van den Berg, A. B. & CDNA 1993. Zeldzame en schaarse vogels in Nederland in 1991. *Limosa* 66: 153-160.

de By, R. A. & CDNA 1991. Zeldzame en schaarse vogels in Nederland in 1989. *Limosa* 64: 61-68.

de By, R. A. & de Knijff, P. 1989. Zeldzame en schaarse vogels in Nederland in 1988. *Limosa* 62: 195-206.

de By, R. A. & Winkelman, J. E. 1987. Zeldzame en schaarse vogels in Nederland in 1986. *Limosa* 60: 195-204.

Byars, T. & Galbraith, H. 1980. Cape May Warbler: new to Britain and Ireland. *British Birds* 73: 2-5.

Byers, C., Olsson, U. & Curson, J. 1995. *Buntings and sparrows: a guide to the buntings and North American sparrows*. Mountfield.

CAF 2006. Décisions prises par la Commission de l'Avifaune Française en 2004-2005. *Ornithos* 13: 244-257.

Campey, R. 1997. The Veery on Lundy. *Birding World* 10: 184.

Campey, R. & Mortimer, K. 1990. Ancient Murrelet on Lundy - A new Western Palearctic bird. *Birding World* 3: 211-212.

Catley, G. 2009. A Soft-plumaged Petrel in Arctic Norway – the first record for the North Atlantic. *Birding World* 22: 249-252.

Cederroth, C. 1995. Sällsynta fåglar i Sverige 1994 - rapport från SOF:s raritetskommitté. *Vår Fågelvärld Supplement* 22: 125-146.

Cederroth, C. 1998. Sällsynta fåglar i Sverige 1997 - rapport från SOF:s raritetskommitté. *Vår Fågelvärld Supplement* 30: 141-169.

Cederroth, C. 2000. Sällsynta fåglar i Sverige 1999 - rapport från SOF:s raritetskommitté. *Vår Fågelvärld Supplement* 33: 137-163.

Cederroth, C. 2002. Sällsynta fåglar i Sverige 2001 - rapport från SOF:s raritetskommitté. *Vår Fågelvärld Supplement* 37: 123-145.

Cederroth, C. 2004. Sällsynta fåglar i Sverige 2003 - rapport från SOF:s raritetskommitté. *Vår Fågelvärld Supplement* 42: 171-193.

Chantler, P. & Driessens, G. 2000. *Swifts: a guide to the Swifts and Treeswifts of the World*. Second edition. Mountfield.

Charlton, T. 1991. The position of Lark Sparrow on the British list. *Birding World* 4: 156-159.

Charlton, T. D. 1995. Lark Sparrow in Suffolk: new to the Western Palearctic. *British Birds* 88: 395-400.

Cheke, R. A., Mann, C. F. & Allen, R. 2001. *Sunbirds*. London.

Christensen, N. H. 1960. Brun Fluesnapper (*Muscicapa latirostris* Raffles) ved Blåvand efterår 1959. *Dansk Orn Foren Tidsskr* 54: 36-40.

Claeys, G., De Bruyker, H. & Geers, V. 1989. Eerste waarneming in Belgie en West-Europa van een Indische Ralreiger *Ardeola grayii*. *Oriolus* 55: 73-75.

Clancey, P. A. 1950. American Bald Eagle in Yorkshire. *British Birds* 43: 339.

Clarke, T. 1999. Autumn 1998 on the Azores. *Birding World* 12: 205-212.

Clarke, T. 2006. *Birds of the Atlantic Islands*. London.

Clement, P. & Hathway, R. 2000. *Thrushes*. London.

Clements, A. 1990. A record of Banded Martin *Riparia cincta* from Egypt. *Sandgrouse* 12: 55-56.

Collett, T. 2009. The Glaucous-winged Gull in Cleveland – the second British record. *Birding World* 22: 14-21.

Cornish, T. 1887. Esquimaux Curlew at Scilly. *The Zoologist* 11 (third series): 388.

Corso, A. 2009. Tristan Albatross in the WP: a cautionary note. *Dutch Birding* 31: 307-308.

Corso, A. & Dennis, P. 1998. Amur Falcons in Italy - a new Western Palearctic bird. *Birding World* 11: 259-260.

Costa, H., Bolton, M., Catry, P., Matias, R., Moore, C. C. & Tomé, R. 2000. Aves de ocorrência rara ou accidental em Portugal. Relatório do Comité Português de Raridades referente aos anos de 1997 e 1998. *Pardela* 11: 3-27.

Costa, H., Bolton, M., Matias, R., Moore, C. C. & Tomé, R. 2003. Aves de ocorrência rara ou accidental em Portugal. Relatório do Comité Português de Raridades referente aos anos de 1999, 2001 e 2001. *Anuário Ornitológico* 1: 3-35.

Cox, S. 1988. Northern Mockingbird in Essex. *Birding World* 1: 233-234.

Cox, S. 1996. Northern Mockingbird in Essex. *British Birds* 89: 353-354.

Coyle, S., Grant, T. & Witherall, M. 2004. The Purple Martin on the Outer Hebrides - a new Western Palearctic bird. *Birding World* 17: 381-389.

Coyle, S. P., Grant, T. C. R. & Witherall, M. J. 2007. Purple Martin on Lewis: new to Britain. *British Birds* 100: 143-148.

Cramp, S. (editor) 1985. *The birds of the Western Palearctic*, 4. Oxford.

Cramp, S. (editor) 1988. *The birds of the Western Palearctic*, 5. Oxford.

Cramp, S. (editor) 1992. *The birds of the Western Palearctic*, 6. Oxford.

Cramp, S. & Conder, P. J. 1970. A visit to the oasis of Kufra, spring 1969 [sic]. *Ibis* 112: 261-263.

Cramp, S. & Perrins, C. M. (editors) 1993. *The birds of the Western Palearctic*, 7. Oxford.

Cramp, S. & Perrins, C. M. (editors) 1994a. *The birds of the Western Palearctic*, 8. Oxford.

Cramp, S. & Perrins, C. M. (editors) 1994b. *The birds of the Western Palearctic*, 9. Oxford.

Cramp, S. & Simmons, K. E. L. (editors) 1977. *The birds of the Western Palearctic*, 1. Oxford.

Cramp, S. & Simmons, K. E. L. (editors) 1980. *The birds of the Western Palearctic*, 2. Oxford.

Cramp, S. & Simmons, K. E. L. (editors) 1983. *The birds of the Western Palearctic*, 3. Oxford.

Crochet, P. A. & Spaans, B. 2008. Spur-winged Geese at Banc d'Arguin, Mauritania, in December 2004. *Dutch Birding* 30: 101-102.

Croft, K. 2001. The Grey Catbird on Anglesey. *Birding World* 14: 424-425.

Croft, K. 2004. Grey Catbird on Anglesey: new to Britain. *British Birds* 97: 630-632.

Croft, K. & Davies, A. 1987. An American Flycatcher. *Twitching* 1: 93-94.

Cudworth, J. 1979. Rufous-sided Towhee in North Humberside. *British Birds* 72: 291-293.

Curson, J., Quinn, D. & Beadle, D. 1994. *New World Warblers*. London.

Curtis, W. F. 1993. Yellow-nosed Albatross *Diomedea chlororhynchos* off Cornwall. *Sea Swallow* 42: 63-66.

Cyprus Ornithological Society (1957) 1983. *Annual Report*, 1982. Nicosia.

Dannenberg, R. 1983. Capped Petrel in northeastern North Atlantic Ocean in february 1980. *Dutch Birding* 5: 85-86.

Davidson, P. & Kirwan, G. M. 1998. Around the region. *Sandgrouse* 20: 76-80.

Davydovich, L. I. & Gorban, I. M. 1990. Zheltobrovaya ovsyanka – novy vid v avifaune Ukrainy. *Ornithologia* 24: 147.

Dawson, I. & Allsopp, K. 1986. Recent reports. *British Birds* 79: 1-17.

De Faveri, A., Baccetti, N. & Arcamone, E. 1998. Striped Crake at Livorno, Italy, in January 1997. *Dutch Birding* 20: 172-174.

Dekker, R. W. R. J. 1981. Black Heron in Egypt in August 1980. *Dutch Birding* 3: 80.

Delin, H. 1987. Sällsynta fåglar i Sverige 1985 – rapport från SOF:s raritetskommitté. *Vår Fågelvärld* 46: 309-329.

Dennis, J. V. 1981. A summary of banded North American birds encountered in Europe. *North American Bird Bander* 6: 88-96.

Dennis, M. 1996. The Redhead in Nottingham - a new Western Palearctic bird. *Birding World* 9: 93-97.

Dennis, M. C. 1998. Redhead in Nottinghamshire: new to Britain and Ireland. *British Birds* 91: 149-154.

Deutsche Seltenheitenkommission 1997. Seltene Vogelarten in Deutschland 1995. *Limicola* 11: 153-208.

Deutsche Seltenheitenkommission 2002. Seltene Vogelarten in Deutschland 1998. *Limicola* 16: 113-184.

Deutsche Seltenheitenkommission 2006. Seltene Vogelarten in Deutschland 2000. *Limicola* 20: 281-353.

Deutsche Seltenheitenkommission 2008. Seltene Vogelarten in Deutschland von 2001 bis 2005. *Limicola* 22: 249-339.

Deutsche Seltenheitenkommission 2009. Seltene Vogelarten in Deutschland von 2006 bis 2008. *Limicola* 23: 257-334.

Dierschke, J. 1998. Der Grünwaldsänger *Dendroica virens* von Helgoland. *Ornithologischer Jahresbericht Helgoland* 8: 85-87.

Díes, J. I., Lorenzo, J. A., Gutiérrez, R., García, E., Gorospe, G., Martí-Aledo, J., Gutiérrez, P. & Vidal, C. 2007. Observaciones de aves rares en España, 2005. *Ardeola* 54: 405-446.

Díes, J. I., Lorenzo, J. A., Gutiérrez, R., García, E., Gorospe, G., Martí-Aledo, J., Gutiérrez, P., Vidal, C., Sales, S. & López-Valasco, D. 2008. Observaciones de aves rares en España, 2006. *Ardeola* 55: 259-287.

Díes, J. I., Lorenzo, J. A., Gutiérrez, R., García, E., Gorospe, G., Martí-Aledo, J., Gutiérrez, P., Vidal, C., Sales, S. & López-Valasco, D. 2009. Observaciones de aves rares en España, 2007. *Ardeola* 56: 309-344.

Dix, T. 1992. The Hooded Warbler on St Kilda. *Birding World* 5: 380-381.

Dixey, A. E., Ferguson, A., Heywood, R. & Taylor, A. R. 1981. Aleutian Tern: new to the western Palearctic. *British Birds* 74: 411-416.

Doherty, P. 1992. Golden-winged Warbler: new to the Western Palearctic. *British Birds* 85: 595-600.

Donark, T. 1953. Falkenes udbredselse og forekomst i Danmark III: Lille Tårnfalk, Tårnfalk og Amerikansk Tårnfalk. *Dansk Orn Foren Tidsskr* 47: 104-122.

Donnelly, P. J. 1993. Yellow-browed Bunting in Orkney. *British Birds* 86: 411-414.

Douhan, B. 1989. Glasögonflugsnappare *Muscicapa dauurica* för första gången anträffad i Sverige. *Vår Fågelvärld* 48: 123-126.

Dowdall, J. F. 1995. Philadelphia Vireo: new to the Western Palearctic. *British Birds* 88: 474-477.

Drost, R. 1933. Turdus unicolor Tickell aus dem Himalaya auf Helgoland. *Ornithologische Monatsberichte* 4: 22-23.

Dubois, P. J. & CHN 1997. Les oiseaux rares en France en 1996. *Ornithos* 4: 141-164.

Dubois, P. J., Le Maréchal, P., Olioso, G. & Yésou, P. 2008. *Nouvel inventaire des oiseaux de France*. Paris.

Dubois, P. & Seitre, R. 1997. Herald Petrel: a new species for the Western Palearctic. *Birding World* 10: 456-459.

Dufourny, H. 1997. Rapport de la Commission d'Homologation. Année 1994. *Aves* 34: 73-96.

Dufourny, H. 1999. White-tailed Tropicbird in Cape Verde Islands in February 1999. *Dutch Birding* 21: 254-255.

Dufourny, H. 2006. First record of Long-tailed Shrike *Lanius schach* for Jordan. *Sandgrouse* 28: 73-75.

Dukes, P. 1987. A new British bird - Wood Thrush on St Agnes. *Twitching* 1: 299-300.

Dukes, P. A. 1995. Wood Thrush in Scilly: new to Britain and Ireland. *British Birds* 88: 133-135.

Dunnett, J. B. 1992. Long-toed Stint: new to Britain and Ireland. *British Birds* 85: 431-436.

Dupuy, A. 1969. Catalogue ornithologique du Sahara algérien. *L'Oiseau et RFO* 39: 140-160 & 225-241.

Durand, A. L. 1963. A remarkable fall of American land-birds on the "Mauretania", New York to Southampton, October 1962. *British Birds* 56: 157-164.

Dybbro, T. 1978. *Oversigt over Danmarks fugle 1978*. København.

Dymond, J. N. & Rarities Committee 1976. Report on rare birds in Great Britain in 1975. *British Birds* 69: 321-368.

Ebels, E. B. 1991. Spotlijster op Schiermonnikoog in oktober 1988. *Dutch Birding* 13: 86-89.

Ebels, E. B. 2004. Probable escapes in the Netherlands: part 2. *Dutch Birding* 26: 305-314.

Ebels, E. B. & van den Bergh, L. M. J. 2007. Geelkoptroepiaal op Terschelling in juli 1982. *Dutch Birding* 29: 159-162.

Edholm, M., Granberg, B. & Lundberg, A. 1980. Rostskagtrasten *Catharus fuscescens* funnen i Uppland, en ny art för Sverige. *Vår Fågelvärld* 39: 137-138.

Editors 1972. Sandhill Crane in Co. Cork in 1905. *British Birds* 65: 427.

Edwards, K. D. & Osborne, K. C. 1972. Hooded Warbler in the Isles of Scilly: a species new to Britain and Ireland. *British Birds* 65: 203-205.

Eischer, K. 1995. The Thick-billed Warbler in Finland. *Birding World* 8: 10-11.

Elias, G., Costa, H., Matias, R., Moore, C. C. & Tomé, R. 2004. Aves de ocorrência rara ou accidental em Portugal. Relatório do Comité Português de Raridades referente ao ano de 2002. *Anuário Ornitológico* 2: 1-20.

Elkins, N. 1979. Nearctic landbirds in Britain and Ireland: a meteorological analysis. *British Birds* 72: 417-433.

Elkins, N. 1988. Recent transatlantic vagrancy of landbirds and shorebirds. *British Birds* 81: 484-491.

Elkins, N. 1999. Recent records of Nearctic landbirds in Britain and Ireland. *British Birds* 92: 83-95.

Elkins, N. 2008. Further thoughts on the transatlantic vagrancy of landbirds to Britain & Ireland. *British Birds* 101: 458-477.

Ellis, P. 1991. The Sandhill Crane in Shetland. *Birding World* 4: 322-323.

Ellis, P. M. & Riddiford, N. J. 1992. Savannah Sparrow in Shetland: second record for the Western Palearctic. *British Birds* 85: 561-564.

Elmberg, J. 1992. Sällsynta fåglar i Sverige 1991 - rapport från SOF:s raritetskommitté. *Vår Fågelvärld* 51 (7-8): 17-32.

Enright, S. D. 1995. Magnolia Warbler in Scilly: new to Britain and Ireland. *British Birds* 88: 107-108.

Eriksen, J., Sargeant, D. E. & Victor, R. 2003. *Oman Bird List: The official list of the birds of the Sultanate of Oman*. Edition 6. Sultan Qaboos University, Muscat.

Eriksen, J., Sargeant, D. E. & Victor, R. 2010. *Additions and corrections to Oman Bird List*, Edition 6. Website: http://www.birdsoman.com/obl6-update.html.

Evans, L. G. R. E. 1994. *Rare birds in Britain 1800-1990*. Little Chalfont.

Everett, M. J. 1992. Wedge-tailed Shearwater *Puffinus pacificus* at Port Said. *Courser* 3: 52-54.

Ferguson, D. 1997. Bay-breasted Warbler in Cornwall: new to Britain and Ireland. *British Birds* 90: 444-449.

Ferrand, Y., Gossmann, F., Guillaud-Rollin, Y. & Jiguet, F. 2008. Première mention de la Bécasse d'Amérique *Scolopax minor* pour la France et le Paléarctique occidental. *Ornithos* 15: 128-131.

Fleet, D. 1982. Brown Flycatcher in BRD in August 1982. *Dutch Birding* 4: 97-98.

Fominykh, M. A. 2007. [On rare birds in the Republic of Bashkortostan]. Materialy k rasprostraneniyu ptits na Urale, v Priural'ye I Zapadnoy Sibiri 74-75.

Forrester, R. W., Andrews, I. J., McInerny, C. J., Murray, R. D., McGowan, R. Y., Zonfrillo, B., Betts, M. W., Jardine, D. C. & Grundy, D. S. (editors) 2007. *The Birds of Scotland*. Aberlady.

Foster, B. 2006. *The Kuwait Bird Report 2005*. Website: http://alsirhan.com/Birds/The_Kuwait_Bird_Report_2005.htm.

Foster, K. 2000. The Siberian Blue Robin in Suffolk - a new British bird. *Birding World* 13: 412-414.

Foster, K. 2006. Siberian Blue Robin at Minsmere: new to Britain. *British Birds* 99: 517-520.

Fouquet, M. 1996. Première mention du Choucas de Daourie *Corvus dauuricus* en France. *Ornithos* 3: 145-146.

Fracasso, F., Baccetti, N. & Serra, L. 2009. La lista CISO-COI degli Uccelli italiani – Parte prima: liste A, B e C. *Avocetta* 33: 5-24.

Fraser, P. A. & Rarities Committee 2007. Report on rare birds in Great Britain in 2006. *British Birds* 100: 694-754.

Fraser, P. A., Rogers, M. J. & Rarities Committee 2007a. Report on rare birds in Great Britain in 2005 part 1: non-passerines. *British Birds* 100: 16-61.

Fraser, P. A., Rogers, M. J. & Rarities Committee 2007b. Report on rare birds in Great Britain in 2005 part 2: passerines. *British Birds* 100: 74-104.

Frémont, J.-Y. & CHN 1999. Les oiseaux rares en France en 1998. *Ornithos* 6: 145-172.

Frémont, J.-Y. & CHN 2003. Les oiseaux rares en France en 2001. *Ornithos* 10: 49-83.

Frémont, J.-Y. & CHN 2005. Les oiseaux rares en France en 2003. *Ornithos* 12: 2-45.

Fry, C. H., Fry, K. & Harris, A. 1992. *Kingfishers, Bee-eaters & Rollers*. London.

Gantlett, S. 1993. The Pacific Swift in Norfolk. *Birding World* 6: 190-191.

Gantlett, S. 1996. 1995: the Western Palearctic year. *Birding World* 9: 21-36.

Gantlett, S. 1999. 1998: the Western Palearctic year. *Birding World* 12: 13-30.

Gantlett, S. 2000. 1999: the Western Palearctic year. *Birding World* 13: 15-32.

REFERENCES

Gantlett, S. 2008. 2007: the Western Palearctic year. *Birding World* 21: 29-42.

Gantlett, S. 2007. 2006: the Western Palearctic year. *Birding World* 20: 26-43.

Gantlett, S. 2009. 2008: the Western Palearctic year. *Birding World* 22: 28-42.

Gantlett, S. 2010. 2009: the Western Palearctic year. *Birding World* 23: 22-40.

Gantlett, S. & Pym, T. 2007. The Atlantic Yellow-nosed Albatross from Somerset to Lincolnshire - a new British bird. *Birding World* 20: 279-295.

Garðarsson, A. 1997. Korpönd að vestan. *Bliki* 18: 65-67.

Garner, M., Lewington, I. & Rosenberg, G. 2004. Stejneger's Scoter in the Western Palearctic and North America. *Birding World* 17: 337-347.

Gätke, H. 1900. *Die Vogelwarte Helgoland*. Second edition. Braunschweig.

Gavrilov, E. I. & Gavrilov, A. E. 2000. Possible ringing recoveries of Relict Gull in Bulgaria and Turkey. *Dutch Birding* 22: 219-221.

Ghigi, A. 1932. Spedizione scientifica all'oasi di Cufra (marzo-luglio 1931) – ucelli. *Annali del Museo Civico di Storia Naturale di Genova* 55: 268-292.

Gibbins, C. & Hackett, P. 2009. The Slaty-backed Gull in Latvia – a second first for the Western Palearctic. *Birding World* 22: 148-150.

Gichuki, C. M. & Pearson, D. J. 1987. A Kenya record of the Shy Albatross *Diomedea cauta*. *Scopus* 11: 44.

Gill, F. & Donsker, D. (editors) 2010. IOC World Bird Names (version 2.6). Website http://www.worldbirdnames.org.

Goodman, S. M. & Meininger, P. L. 1989. *The birds of Egypt*. Oxford.

Goodman, S. M. & Storer, R. W. 1987. The seabirds of the Egyptian Red Sea and adjacent waters, with notes on selected Ciconiiformes. *Gerfaut* 77: 109-145.

Grace, K. & Lancaster, A. A. K. 1990. Ovenbird on Dursey Island, Co. Cork. *Irish Birding News* 1: 53-55.

Le Grand, G. 1983. Bilan des Observations sur les Oiseaux d´origine Néarctique Effectuées aux Açores (Jusqu'en Janvier 1983). *Arquipélago* 4: 73-83.

Le Grand, G. 1993. *Recherches sur l'écologie des vertébrés terrestres de l'archipel des Açores*. Thése de Doctorate de l´École Pratique de Hautes Études, University of Montpellier. Montpellier.

Grant, P. 1987. The Co. Kerry Bald Eagle. *Twitching* 1: 379-380.

Grantham, M. 2005. An Amur Falcon in France, North Yorkshire and Dumfries & Galloway – a belated first for the Western Palearctic? *Birding World* 18: 289.

Gregory, G. 2005. *Birds of Kuwait*. Skegness.

Grieve, A. 1987. Hudsonian Godwit: new to the Western Palearctic. *British Birds* 80: 466-473.

Grieve, A. 1992. First record of Thick-billed Warbler *Acrocephalus aedon* in Egypt. *Sandgrouse* 14: 123-124.

Griffiths, E. 1996. Northern Mockingbird in Cornwall. *British Birds* 89: 353.

Grošelj, P. 1988. *Emberiza rutila*, brezimni gost iz vzhodne Azije / *Emberiza rutila*, an anonymous guest from East Asia. *Acrocephalus* 37-38: 64-66.

Grygoruk, G. & Tumiel, T. 2006a. The Rufous-tailed Robin in Poland. *Birding World* 19: 17.

Grygoruk, G. & Tumiel, T. 2006b. Pierwsze w Polsce i drugie w zachodniej Palearktyce stwierdzenie słowika syberyjskiego *Luscinia sibilans*. *Notatki Ornitologiczne* 47: 192-194.

Gustad, J. R. 1994. Sjeldne fugler i Norge i 1992. *Vår Fuglefauna* 17: 259-278.

Gustad, J. R. 1995a. Sjeldne fugler i Norge i 1993 og 1994. *Vår Fuglefauna* 18: 259-302.

Gustad, J. R. 1995b. The Norwegian Brown-headed Cowbird. *British Birds* 88: 377.

Gutiérrez, R. & Lorenzo, J. A. 1999. The waterthrush on the Canary Islands in 1991. *Birding World* 12: 294-295.

Gutiérrez, R. 2008. The Fork-tailed Flycatcher in Spain – a new Western Palearctic bird. *Birding World* 21: 325-328.

Haas, M. 2009. Tristan Albatross collected in Sicily, Italy, in October 1957. *Dutch Birding* 31: 180.

Haas, M. & Crochet, P.-A. 2009. Western Palearctic list updates: Cape Petrel. *Dutch Birding* 31: 24-28.

Haas, M., Crochet, P.-A. & Lamarche, B. 2010. Western Palearctic list updates: African Palm Swift. *Dutch Birding* 32: 131-132.

Haas, M., Crochet, P.-A. & Shigeta, Y. in prep. Western Palearctic list updates: Von Schrenck's Bittern. *Dutch Birding*.

Haas, M., Legrand, V. & Monticelli, D. 2010. Three-banded Plovers breeding at Aswan, Egypt, in 2009. *Dutch Birding* 32: 126-128.

Haavisto, S. & Strand, A. 2000. The first Lesser Moorhen *Gallinula angulata* in Egypt and the Western Palearctic. *Sandgrouse* 22: 137-139.

Hadarics, T. & Schmidt, A. 1997. Az MME NB közleménye a Fehértón megfigyelt és korábban hosszúfarkú gébicsnek (*Lanius schach*) határozott madárral kapcsolatban. *Túzok* 2: 128.

Haftorn, S. 1971. *Norges Fugler*. Oslo.

Hanafin, M. 2006. The Canada Warbler in County Clare - the second for the Western Palearctic. *Birding World* 19: 429-434.

Handrinos, G. & Akriotis, A. 1996. *The Birds of Greece*. London.

Hansen, E., Hansen, P. S. & Nielsen, B. P. 1974. Rapport fra sjældenhedsudvalget for 1973. *Dansk Orn Foren Tidsskr* 68: 138-144.

Harding, J. 2004. The Ovenbird on the Isles of Scilly. *Birding World* 17: 424-425.

Harrop, H. 2003a. A Masked Booby in the Bay of Biscay. *Birding World* 16: 388-390.

Harrop, H. 2003b. Un Fou masqué *Sula dactylatra* dans le golfe de Gascogne. *Ornithos* 10: 294-295.

Harting, J. E. 1879. Esquimaux Curlew in Aberdeenshire. The Zoologist 3 (third series): 135.

den Hartog, J. C. 1987. A record of a Red-footed Booby *Sula sula* (L.) from the Cape Verde Islands, with a review of the status of this species in the South Atlantic Ocean. *Zoologische Mededelingen* 61: 405-419.

Harvey, P. 1992. The Brown Flycatcher on Fair Isle – a new British bird. *Birding World* 5: 252-255.

Harvey, P. 2001. The Thick-billed Warbler on Shetland. *Birding World* 14: 372-373.

Harvey, P. 2010. Brown Flycatcher on Fair Isle: new to Britain. *British Birds* 103: 651-657.

Harvie Brown, J. A. 1880. Rare birds and the autumnal migration. *The Zoologist* 4 (third series): 485-486.

Hatton, D. & Varney, P. 1989. Red-breasted Nuthatch in Norfolk - A new British Bird. *Birding World* 2: 354-356.

Hazevoet, C. J. 1985. Black Noddy in Mauritania in December 1984. *Dutch Birding* 7: 25-26.

Hazevoet, C. J. 1995. *The birds of the Cape Verde Islands*. BOU Check-list 13. Tring.

Hazevoet, C. J. 1999. Fourth report on birds from the Cape Verde Islands, including notes on conservation and records of 11 taxa new to the archipelago. *Bulletin of the Zoological Museum of the University of Amsterdam* 17: 19-32.

Hazevoet, C. J. 2010. Sixth report on birds from the Cape Verde Islands, including records of 25 taxa new to the archipelago. *Zoologia Caboverdiana* 1 (1): 3-44.

Heim de Balsac, H. & Mayaud, N. 1962. *Les Oiseaux du Nord-Ouest de l'Afrique*. Paris.

Heinroth, O. 1908. Eine Katzendrossel *Galeoscoptes carolinensis* L. bei Anklam beobachtet. *Ornithologische Monatsberichte* 16: 143-144.

Hellenic Rarities Committee 2009. Annual Report – 2008. Website http://rarities.ornithologiki.gr/gr/eaop/annual_reports.htm.

Hellström, M. & Strid, T. 2008. Fågelrapport 2007. *Vår Fågelvärld* Supplement 48: 43-156.

Hickman, D. J. D. 1995. Tree Swallow in Scilly: new to the Western Palearctic. *British Birds* 88: 381-384.

Higson, P. 2009. The Sandhill Crane on Orkney. *Birding World* 22: 376-378.

Hirschfeld, E. 1987. Sällsynta fåglar i Sverige 1986 – rapport från SOF:s raritetskommitté. *Vår Fågelvärld* 46: 441-456.

Hjartarson, G., Rikarðsson, R. & Kolbeinsson, Y. 2003. Þrjár nýjar skríkjutegundir berast til Íslands. *Bliki* 23: 51-54.

Hoath, R. 2000. The first Three-banded Plover *Charadrius tricollaris* in Egypt and the Western Palearctic. *Sandgrouse* 22: 67-68.

Holden, D. & Bilton, D. 2009. The Eastern Crowned Warbler in County Durham – a new British Bird. *Birding World* 22: 417-419.

Holliday, S. T. 1990. List of unpublished bird records of significance for Gibraltar to 31 December 1986. *Alectoris* 7: 49-57.

Holman, D. 1990. Britain's first Yellow-browed bunting. *British Birds* 83: 430-432.

Hoogendoorn, W. 1991. Record of Brown-headed Gull in Israel in May 1985. *Dutch Birding* 13: 104-106.

Hopkins, D., Stone, D. & Rylands, K. 2006. The Long-billed Murrelet in Devon - a new British bird. *Birding World* 19: 457-464.

Hørring, R. 1933. *Aethia cristatella* (Pallas) skudt ved Island. *Dansk Orn Foren Tidsskr* 27: 103-105.

Hovette, C. 1972. Nouvelles acquisitions avifaunistiques de la Camargue. *Alauda* 40: 343-352.

Hovette, C. & Kowalski, H. 1972. Observations de Camargue. Héron mélanocéphale. *Alauda* 40: 397.

Howell, S. 2002. A Black-capped Petrel off the Bay of Biscay: the fourth for the Western Palearctic. *Birding World* 15: 219-220.

Høyland, B. O., Heggland, H. & Mjøs, A. T. 2001. Sjeldne fugler i Norge i 1997 og 1998. *Vår Fuglefauna* Supplement 4: 4-31.

del Hoyo, J., Elliot, A. & Christie, D. A. (editors) 2004. *Handbook of the birds of the world* 9. Barcelona.

del Hoyo, J., Elliot, A. & Christie, D. A. (editors) 2005. *Handbook of the birds of the world* 10. Barcelona.

del Hoyo, J., Elliot, A. & Christie, D. A. (editors) 2005. *Handbook of the birds of the world* 11. Barcelona.

del Hoyo, J., Elliot, A. & Christie, D. A. (editors) 2005. *Handbook of the birds of the world* 13. Barcelona.

del Hoyo, J., Elliot, A. & Christie, D. A. (editors) 2005. *Handbook of the birds of the world* 14. Barcelona.

del Hoyo, J., Elliot, A. & Christie, D. A. (editors) 2005. *Handbook of the birds of the world* 15. Barcelona.

del Hoyo, J., Elliot, A. & Sargatal, J. (editors) 1992. *Handbook of the birds of the world* 1. Barcelona.

del Hoyo, J., Elliot, A. & Sargatal, J. (editors) 1994. *Handbook of the birds of the world* 2. Barcelona.

del Hoyo, J., Elliot, A. & Sargatal, J. (editors) 1994. *Handbook of the birds of the world* 3. Barcelona.

del Hoyo, J., Elliot, A. & Sargatal, J. (editors) 1994. *Handbook of the birds of the world* 4. Barcelona.

del Hoyo, J., Elliot, A. & Sargatal, J. (editors) 1994. *Handbook of the birds of the world* 5. Barcelona.

del Hoyo, J., Elliot, A. & Sargatal, J. (editors) 1994. *Handbook of the birds of the world* 6. Barcelona.

del Hoyo, J., Elliot, A. & Sargatal, J. (editors) 1994. *Handbook of the birds of the world* 7. Barcelona.

Hudson, N. & Rarities Committee 2008. Report on rare birds in Great Britain in 2007. *British Birds* 101: 516-577.

Hudson, N. & Rarities Committee 2009. Report on rare birds in Great Britain in 2008. *British Birds* 102: 528-601.

Hudson, N. & Rarities Committee 2010. Report on rare birds in Great Britain in 2009. *British Birds* 103: 562-638.

Hunt, D. B. 1979. Yellow-bellied Sapsucker: new to Britain and Ireland. *British Birds* 72: 410-414.

Hutchinson, C. D. 1989. *Birds in Ireland*. Calton.

van IJzendoorn, E. J., van der Laan, J. & CDNA 1996. Herziening Nederlandse Avifaunistische Lijst 1800-1979: tweede fase. *Dutch Birding* 18: 157-202.

Incledon, C. S. C. 1968. Brown Thrasher in Dorset: a species new to Britain and Ireland. *British Birds* 61: 550-553.

Ingram, C. 1929. Première capture en Europe et en France de *Locustella fasciolata* (Gray). *Alauda* 1: 292.

Ingram, C. 1930. Première capture en Europe et en France de *Locustella fasciolata* (Gray). *L'Oiseau et RFO* 11: 57.

Irish Rare Birds Committee 1998. *Checklist of the Birds of Ireland*. Monkstown.

Isenmann, P. 2007. *The birds of the Banc d'Arguink*. Montpellier.

Isenmann, P. & Moali, A. 2000. *The birds of Algeria*. Paris.

Israeli Rarities and Distribution Committee 2002. Bulletin 2:02 on Rare Birds in Israel. Website http://www.israbirding.com/irdc/bulletins/bulletin_2.2/.

James, M. 1988. Daurian Redstart in Fife – A New Western Palearctic Bird. *Birding World* 1: 162-163.

Jännes, H. 1995. Paksunokkakerttunen - kerttuseksi helppo tuntea. *Alula* 1: 18-20.

Janni, O. & Fracasso, G. 2009. Commissione Ornitologica Italiana (COI). Report 22. *Avocetta* 33: 116-146.

Janni, O., Fracasso, G. 2009. Commissione Ornitologica Italiana (COI) - Report 22. *Avocetta* 33: 116-146.

Jany, E. 1963. Salma Kabir - Kufra -Djabal al Uwenat. Ein Reisebericht aus der östlichen Sahara. *Die Erde* 94: 334-362.

Jara, J., Costa, H., Elias, G. Matias, R., Moore, C. C., Noivo, C. & Tipper, R. 2008. Aves de ocorrência rara ou accidental em Portugal. Relatório do Comité Português de Raridades referente aos anos de 2006 e 2007. *Anuário Ornitológico* 6: 1-45.

Jara, J., Costa, H., Elias, G., Matias, R., Moore, C. C. & Tomé, R. 2007. Aves de ocorrência rara ou accidental em Portugal. Relatório do Comité Português de Raridades referente ao ano de 2005. *Anuário Ornitológico* 5: 1-34.

Jiguet, F. 2006. Identification of non-breeding Sqacco, Indian and Chinese Pond Herons. *Alula* 12: 114-119.

Jiguet, F. 2007. Première mention française du Bruant à sourcils jaunes *Emberiza chrysophrys*. *Ornithos* 14: 172-175.

Jiguet, F. & CAF 2004. Décisions récentes prises par la Commission de l'Avifaune Française. *Ornithos* 11: 230-245.

Jiguet, F. & CAF 2007. Première mention pour la France de la Macreuse á aisles blanches sibérienne. *Ornithos* 14: 38-42.

REFERENCES

Jiguet, F., Crochet, P.-A., Dubois, P. J., Le Maréchal, P., Pons, J.-M. & Yésou, P. 2007. Décisions prises par la Commission de l'Avifaune Française en 2006-2007. *Ornithos* 14: 108-115.

Jiguet, F., Crochet, P-A, Dubois, P J, Pons, J-M, Yésou, P & Le Maréchal, P 2009. Décisions prises par la Commission de l'Avifaune Française en 2008-2009. *Ornithos* 16: 382-393.

Jiguet, F. & Defos du Rau, C. 2004. A Cape Gull in Paris - a new European bird. *Birding World* 17: 62-70.

de Juana, E. 2006. *Aves raras de España. Un catálogo de las especies de presentación ocasional*. Bellaterra.

de Juana, E. & Comité de Rarezas de la SEO 1998. Observaciones de Aves Raras en España, año 1996. *Ardeola* 45: 97-116.

de Juana, E. & Comité de Rarezas de la SEO 2002. Observaciones de Aves Raras en España, año 2000. *Ardeola* 49: 141-171.

de Juana, E. & Comité de Rarezas de la SEO 2003. Observaciones de Aves Raras en España, año 2001. *Ardeola* 50: 123-149.

de Juana, E. & Comité de Rarezas de la SEO 2004. Observaciones de Aves Raras en España, año 2002. *Ardeola* 51: 513-539.

de Juana, E. & Comité de Rarezas de la SEO 2005. Observaciones de Aves Raras en España, año 2003. *Ardeola* 52: 185-206.

de Juana, E & Comité de Rarezas de la SEO 2006. Observaciones de Aves Raras en España, año 2004. *Ardeola* 53: 163-190.

de Juana, E. & Comité Ibérico de Rarezas de la SEO 1988. Observaciones Homologadas de Aves Raras en España. Informe de 1986. *Ardeola* 35: 167-174.

Johnsen, S. 1937. Brille-efuglen [*Arctonetta fischeri* (Brandt)] ny for Norges og Europas fauna. Bergens Museums Årbok 1937. *Naturvidenskapelig rekke* Nr. 3. Bergen.

Junge, G. C. A. & Koch, J. C. 1938. De rosse gors, *Emberiza rutila* Pall., in Nederland. *Limosa* 11: 1-6.

Kainady, P. V. G. 1976. The Indian Pygmy Goose *Nettapus coromandelianus* in Basrah, Iraq. *Bulletin of Basrah Natural History Museum* 3: 107-109.

Kennerley, P. R. & Pr s-Jones, R. P. 2006. Occurrences of Gray's Grasshopper Warbler in Europe, including a further case of Meinertzhagen fraud. *British Birds* 99: 506-516.

King, J. M. B. 1990. Veery in Devon. *British Birds* 83: 284-287.

Kirwan, G. M., Boyla, K., Castell, P., Demirci, B., Özen, M., Welch, H. & Marlow, T. 2008. *The birds of Turkey*. London.

Kirwan, G. M. & Martins, R. P. 1994. Turkey Bird Report 1987-91. *Sandgrouse* 16: 77-117.

Kitson, A. R. & Robertson, I. S. 1983. Yellow-browed Bunting: new to Britain and Ireland. *British Birds* 76: 217-225.

Kivivuori, H., Lehikoinen, A., Lehikoinen, P. && Lindén, A. 2008. Swinhoe's Snipe at Tohmajärvi in summer 2008. *Alula* 14: 124-131.

Knaus, P. & Balzari, C. A. 1999. Seltene Vogelarten und ungewöhnliche Vogelbeobachtungen in der Schweiz im Jahre 1998. *Der Ornithologische Beobachter* 96: 157-182.

Knaus, P., Balzari, C. A. & Maumary, L. 2000. Oiseaux rares et observations inhabituelles en Suisse en 1998. *Nos Oiseaux* 47: 29-49.

Knox, A. G. 1993a. Daurian Redstart in Scotland: captive origin and the British List. *British Birds* 86: 359-366.

Knox, A. G. 1993b. Richard Meinertzhagen—a case of fraud examined. *Ibis* 135: 320-325.

Kolbeinsson, Y. 2004. The Varied Thrush in Iceland. *Birding World* 17: 206-208.

Kolbeinsson, Y. 2005. Tvær nýjar andategundir á Íslenska listann. *Bliki* 26: 57-60.

Kolbeinsson, Y. 2006. Grænskríkja finnst á Íslandi. *Bliki* 27: 69-71.

Kolbeinsson, Y., Arnarson, B. G. & Hilmarsson, J. Ó. 2006. Tveir nýir greipar berast til Íslands og Evrópu. *Bliki* 27: 63-67.

Kolbeinsson, Y., Þráinsson, G. & Pétursson, G. 2001. Sjaldgæfir fuglar á Íslandi 1998. *Bliki* 22: 21-46.

Kolbeinsson, Y., Þráinsson, G. & Pétursson, G. 2003. Sjaldgæfir fuglar á Íslandi 2000. *Bliki* 24: 25-52.

Kolbeinsson, Y., Þráinsson, G. & Pétursson, G. 2004. Sjaldgæfir fuglar á Íslandi 2001. *Bliki* 25: 25-48.

Kolbeinsson, Y., Þráinsson, G. & Pétursson, G. 2005. Sjaldgæfir fuglar á Íslandi 2002. *Bliki* 26: 21-46.

Kolbeinsson, Y., Þráinsson, G. & Pétursson, G. 2006. Sjaldgæfir fuglar á Íslandi 2003. *Bliki* 27: 27-50.

Kolbeinsson, Y., Þráinsson, G. & Pétursson, G. 2007. Sjaldgæfir fuglar á Íslandi 2004. *Bliki* 28: 25-50.

Kolbeinsson, Y., Þráinsson, G. & Pétursson, G. 2008. Sjaldgæfir fuglar á Íslandi 2005. *Bliki* 29: 23-44.

Komisja Faunistyczna 2006. Rzadkie ptaki obserwowane w Polsce w roku 2005. *Notatki Ornitologiczne* 47: 97-124.

Komisja Faunistyczna 2008. Rzadkie ptaki obserwowane w Polsce w roku 2007. *Notatki Ornitologiczne* 49: 81-115.

Kondratyev, A. V. & Zöckler, C. 2009. Mixed pair of Ross's Goose and Barnacle Goose breeding on Kolguev, Russia, in 2006-07. *Dutch Birding* 31: 299-301.

KORC 2008. Kuwait Bird Report 2007. Website: http://www.alsirhan.com/Kuwait_Bird_Report_2007_Final.pdf.

Kristensen, A. B., Frich, A. S., Ortvad, T. E. & Schwalbe, M. 2008. Sjældne fugle i Danmark og Grønland i 2007. *Fugleåret* 2: 117-135.

Kristensen, A. B., Frich, A. S., Ortvad, T. E. & Schwalbe, M. 2009. Sjældne fugle i Danmark og Grønland i 2008. *Fugleåret* 3: 123-144.

Lansdell, J., Lansdell, C., Wilkinson, A. & Gregory, L. 2008. A Lesser Frigatebird in Kuwait – the third Western Palearctic record. *Birding World* 21: 254.

Lawlor, M. 2010. The Pacific Diver in the Channel Islands. *Birding World* 23: 20-21.

Lehikoinen, A., Aalto, T., Nikander, P. J., Normaja, J., Rauste, V. & Velmala, W. 2009. Rariteettikomitean hyväksymät vuoden 2008 harvinaisuushavainnot. *Linnut-Vuosikirja* 2008: 90-103.

Lewington, I., Alström, P. & Colston, P. 1991. *A field guide to the rare birds of Britain and Europe*. London.

Lindholm, T., Aalto, T., Normaja, J., Rauste, V. & Velmala, W. 2008. Rariteettikomitean hyväksymät vuoden 2007 harvinaisuushavainnot. *Linnut-Vuosikirja* 2007: 126-139.

Lindroos, T. 1997a. Rare birds in Finland 1996. *Alula* 4: 160-169.

Lindroos, T. 1997b. Rariteettikomitean hyväksymät havainnot vuodelta 1996. *Linnut* 32 (6): 18-30.

Lobb, M. G. 1983. Didric Cuckoo *Chrysoccocyx caprius* in Cyprus - new to the Palearctic. *Bulletin of the British Ornithologists' Club* 103: 111.

Long, R. 1981a. Review of birds in the Channel Islands, 1951-80. *British Birds* 74: 327-344.

Long, R. 1981b. Catbird in the Channel Islands. *British Birds* 74: 526-527.

Lorenzo, J. A. 2002. The African Crake on Tenerife - a new Western Palearctic bird. *Birding World* 15: 60-61.

Luoto, H., Aalto, T., Lindholm, A., Nikander, P. J., Normaya, J., Soilevaara, K. & Rauste, V. 2004. Rariteettikomitean hyväksymät vuoden 2003 harvinaisuushavainnot. *Linnut-Vuosikirja* 2003: 33-48.

Luoto, H., Aalto, T., Lindholm, A., Normaya, J. & Rauste, V. 2005. Rariteettikomitean hyväksymät vuoden 2004 harvinaisuushavainnot. *Linnut-Vuosikirja* 2004: 73-86.

Luoto, H., Aalto, T., Lindholm, A. & Rauste, V. 2003. Rariteettikomitean hyväksymät vuoden 2002 harvinaisuushavainnot. *Linnut-Vuosikirja* 2002: 63-80.

Madge, S. & Burn, H. 1988. *Wildfowl: an identification guide to the ducks, geese and swans of the world*. London.

Madge, S. C., Heard, G. C., Hutchings, S. C. & Williams, L. P. 1990. Varied Thrush: new to the Western Palearctic. *British Birds* 83: 187-195.

Mahé, E. 1985. *Contribution à l'étude scientifique de la région de Banc d'Arguin*. Montpellier.

Malbrant, R. 1957. Note additionnelle sur les oiseaux du Borkou-Ennedi-Tibesti. *L'Oiseau et RFO* 27: 214-231.

Malczyk, P. & Łukasik, D. 2008. Czwarte stwierdzenie "uhli azjatyckiej" *Melanitta deglandi stejnegeri* w Palearktyce Zachodniej. *Notatki Ornitologiczne* 49: 245-257.

Mansell, D. 2008. The Amur Falcon in East Yorkshire – a new British bird. *Birding World* 21: 432-435.

Martins, R. P. & Hirschfeld, E. 1998. Comments on the limits of the Western Palearctic in Iran and the Arabian peninsula. *Sandgrouse* 20: 108-134.

Massa, B. 1974. La Procellaria del Capo (*Daption capensis* (L.)) è giunta anche nel Mediterraneo. *Rivista Italiana di Ornitologia* 44: 210-212.

Mather, J. H. 2010. Pacific Diver in Yorkshire: new to Britain and the Western Palearctic. *British Birds* 103: 539-545.

Mather, J. R. & Curtis, W. F. 1987. The Barmston Capped Petrel. *British Birds* 80: 284-286.

Matias, R. 2009. Removal of Black Crake *Amaurornis flavirostris* from the Western Palearctic list, and the first record of Lesser Moorhen *Gallinula angulata* for Madeira. *Bulletin of the British Ornithologists' Club* 129 (2): 116-119.

Maumary, L. & Knaus, P. 2000. Marbled Murrelet in Switzerland: a Pacific Ocean auk new to the Western Palearctic. *British Birds* 93: 190-199.

McGeehan, A. & Nash, C. 2009. The Cedar Waxwing in County Galway. *Birding World* 22: 420-423.

McKay, C. R. 1994. Brown-headed Cowbird in Strathclyde: new to Britain and Ireland. *British Birds* 87: 284-288.

McKay, C. R. 2000. Cedar Waxwing in Shetland: new to the Western Palearctic. *British Birds* 93: 580-587.

McLaren, I. A., Lees, A. C., Field, C. & Collins, K. J. 2006. Origins and characteristics of Nearctic landbirds in Britain and Ireland in autumn: a statistical analysis. *Ibis* 148: 707-726.

McShane, C. 1996. Eastern Phoebe in Devon: new to the Western Palearctic. *British Birds* 89: 103-107.

Meek, E. R. 1984. Tennessee Warbler in Orkney. *British Birds* 77: 160-164.

Meeth, P. 1969. Giant Petrel *(Macronectes giganteus)* in West European waters. *Ardea* 57: 92.

Meeth, P. & Meeth, K. 1988. A Shy Albatross off Somalia. *Sea Swallow* 37: 66.

Meijer, A. W. J. 1996. Daurische Kauw in Hollandse kuststreek in mei 1995. *Dutch Birding* 18: 226-231.

Meinertzhagen, R. 1938. Winter in Arctic Lapland. *Ibis* 80: 754-759.

Meinertzhagen, R. H. (editor) 1930. *Nicoll's Birds of Egypt*. London.

Melling, T. 2010. The Eskimo Curlew in Britain. *British Birds* 103: 80-92.

Mellow, B. K. & Maker, P. A. 1981. American Kestrel in Cornwall. *British Birds* 74: 227.

Meyer, K. D. 1995. Swallow-tailed Kite (*Elanoides forficatus*), *The Birds of North America Online* (A. Poole, Ed.). Ithaca: Cornell Lab of Ornithology; Retrieved from the Birds of North America Online: http://bna.birds.cornell.edu/bna/species/138

Michaelsen, J. 1985. Sjeldne fugler i Norge i 1981. *Vår Fuglefauna* 8: 49-52.

Mikkola, K. 1984. Rariteettikomitean hyväksymät vuoden 1983 harvinaisuushavainnot. *Lintumies* 19: 154-167.

Millington, R. 2009. The birding review of 2009. *Birding World* 22: 497-505.

Milne, P. 2005. Fifty-first Irish Bird Report. *Irish Birds* 7: 549-574.

Milne, P. & McAdams D. G. 2008. Irish Rare Bird Report 2006. *Irish Birds* 8: 395-416.

Milne, P. & McAdams D. G. 2009. Irish Rare Bird Report 2007. *Irish Birds* 8: 583-610.

Milne, P., McAdams, D. G. & Dempsey, E. 2002. Forty-eighth Irish Bird Report, 2000. *Irish Birds* 7: 79-110.

Milne, P. & O'Sullivan, O. 1997. Forty-fourth Irish Bird Report, 1996. *Irish Birds* 6: 61-90.

Mitchell, D. & Young, S. 1997. *Photographic handbook to the rare birds of Britain and Ireland*. London.

Mizrachi, R., Perlman, Y., Smith, J. P. & Israeli Rarities and Distribution Committee 2007. *Bulletin* 5:01 on Rare Birds in Israel. Website http://www.israbirding.com/irdc/bulletins/bulletin_5/.

Mjølsnes, K. R., Bunes, V., Olsen, T. A. & Solbakken, K. A. 2006. Sjeldne fugler i Norge i 2004. *Ornis Norvegica* 29: 68-109.

Mjølsnes, K. R., Bunes, V. & Solbakken, K. A. 2005. Sjeldne fugler i Norge i 2003. *Ornis Norvegica* 28: 4-50.

Mjøs, A. T. 2002. Revurdering av eldre funn og endringer på den norske fuglelisten. *Ornis Norvegica* 25: 65-92.

Mjøs, A. T. & Solbakken, K. A. 2001. Sjeldne fugler i Norge 1999 og 2000. *Ornis Norvegica* 24: 3-59.

MME NB 2010. Az MME Nomenclator Bizottság 2006. évi jelentése a Magyarországon ritka madárfajok előfordulásáról. *Aquila* 116-117: 99-114.

Moon, S. J. 1983. Little Whimbrel: new to Britain and Ireland. *British Birds* 76: 438-445.

Moon, S. & Carrington, D. 2002. A Brown Skua in Glamorgan. *Birding World* 15: 387-389.

Morgan, J. & Shirihai, H. 1992. Streaked Shearwaters in Israel - a new Western Palearctic bird. *Birding World* 5: 344-347.

Morozov, V. V. 2004. Displaying Swinhoe's Snipe in eastern European Russia: a new species for Europe. *British Birds* 97: 134-138.

Mullarney, K. & Millington, R. 2008. The Pacific and Black-throated Divers in Pembrokeshire. *Birding World* 21: 63-66.

Müller, H. E. & Lippert, K. 1998. Schwalbenweihe *Elanoides forficatus* auf Fuerteventura - eine neue Art für die Paläarktis. *Limicola* 12: 80-84.

de Naurois, R. 1969. *Peuplements et cycles de reproduction des oiseaux de la côte occidentale d'Afrique*. Mémoires du Muséum National d'Historie Naturelle, Série A, Zoologie. Tome LVI. Paris.

Newell, D. 2008. Recent records of southern skuas in Britain. *British Birds* 101: 439-441.

Newton, A. 1852. Some account of a petrel killed at Southacre, Norfolk, with a description and synonymy. *The Zoologist* 10 (first series): 3691-3698.

Nichols, A. R. 1907. The Canadian Crane in Co. Cork. *The Irish Naturalist* 16: 209-211.

Nicolaisen, H. I. 1987. Ny fugleart for Europa: brunhodekustær på Jomfruland. *Oriolus* 16 (2): 1-2, 22.

Nielsen, H. H. & Thorup, K. 2001. Sjældne fugle i Danmark og Grønland i 2000. *Dansk Orn Foren Tidsskr* 95: 153-166.

von Niethammer, G., Kramer, H. & Wolters, H. E. 1964. *Die Vögel Deutschlands*. Frankfurt am Mein.

Nikander, P. J. & Lindroos, T. 1995. Rariteettikomitean hyväksymät vuoden 1994 harvinaisuushavainnot. *Linnut* 30 (6): 9-18.

Nisbet, I. C. T. 1963a. Three additions to the British and Irish List. Summer Tanager, Baltimore Oriole and Western Sandpiper. *British Birds* 56: 48-58.

Nisbet, I. C. T. 1963b. American passerines in western Europe, 1951-62. *British Birds* 56: 204-217.

Nuovo, G. 2008. Una Sula zamperosse al largo di Capo Verde: seconda osservazione per il Paleartico occidentale! *Quaderni di Birdwatching* 20. Website: http://www.ebnitalia.it/QB/QB020/sula.htm.

Ohlsson, J. 1980. Sällsynta fåglar i Sverige 1978 - rapport från SOF:s raritetskommitté. *Vår Fågelvärld* 39: 35-42.

Ólafsson, E. 1993a. Flækingsfuglar á Íslandi: Hranar og skyldar tegundir, spætur og greipar. *Náttúrufræðingurinn* 62: 83-86.

Ólafsson, E. 1993b. Flækingsfuglar á Íslandi: Tittlingar, græningjar og krakar. *Náttúrufræðingurinn* 63: 87-108.

Ólafsson, E. & Pétursson, G. 1997. Flækingsfuglar á Íslandi: Skríkjur. *Náttúrufræðingurinn* 66: 161-179.

Olsen, K. M. 1988. Sjældne fugle i Danmark og Grønland I 1986 og 1987. *Dansk Ornitologisk Forening Tidsskrift* 82: 81-100.

Olsen, K. M. & Larsson, H. 1997. *Skuas and jaegers: a guide to the skuas and jaegers of the world*. Mountfield.

Olsen, T. A. & Mjølsnes, K. R. 2007. Sjeldne fugler i Norge i 2005. *Ornis Norvegica* 30: 68-115.

Olsen, T. A., Bunes, V., Egeland, Ø., Gullberg, A., Mjølsnes, K. R. & Tveit, B. O. 2010. Sjeldne fugler i Norge i 2008. *Ornis Norvegica* 33: 4-48.

Olsson, C. 1988. Klippkaja för första gången anträffad i Sverige. *Vår Fågelvärld* 47: 197-199.

Onley, D. & Scofield, P. 2007. *Albatrosses, Petrels and Shearwaters of the World*. London.

Orlando, C. 1958. Cattura di un Albatro urlatore *(Diomedea exulans exulans*, Linnaeus) in Sicilia. *Rivista italiana di ornitologia* 28: 101-113.

Ovaa, A., Groenendijk, D., Berlijn, M. & CDNA 2009. Rare birds in the Netherlands in 2008. *Dutch Birding* 31: 331-352.

Ovaa, A., Groenendijk, D., Berlijn, M. & CDNA 2010. Rare birds in the Netherlands in 2009. *Dutch Birding* 32: 363-383.

Ovaa, A., van der Laan, J., Berlijn, M. & CDNA 2008. Rare birds in the Netherlands in 2007. *Dutch Birding* 30: 369-389.

Parker, M. 1990. Pacific Swift: new to the Western Palearctic. *British Birds* 83: 43-46.

Parkin, D. T. & Shaw, K. D. 1994. Asian Brown Flycatcher, Mugimaki Flycatcher and Pallas's Rosefinch. *British Birds* 87: 247-252.

Parrish, R. 1991. The Mugimaki Flycatcher in Humberside – a new Western Palearctic bird. *Birding World* 4: 392-395.

Parrot, J., Phillips, J. & Wood, V. 1987. Tricolored Heron on Azores in October 1985. *Dutch Birding* 9: 17-19.

Paulsen, P. 1925. Seltenheit der Schleswig-holsteinischen Ornis. *Ornithologische Monatsberichte* 50: 163-164.

Peacock, M. 1993. Chestnut-sided Warbler: new to the Western Palearctic. *British Birds* 86: 57-61.

Pedersen, T. & Aspinall, S. (compilers) 2010. EBRC annotated checklist of the birds of the United Arab Emirates. *Sandgrouse* Supplement 3.

Petersen, Æ. 1985. Nýjungar um flækingsfugla á Íslandi. *Bliki* 4: 57-67.

Petersen, Æ. 1989. Fimm nýjar anda- og spörfuglategundir á Íslandi. *Bliki* 8: 56-61.

Pettersson, J., Österberg, J. & Kjellén, N. 1978. Långtåsnäppa *Calidris subminuta* funnen vid Ottenby - en ny art för Europa. *Vår Fågelvärld* 37: 333-338.

Pétursson, G. 1987. Flækingsfuglar á Íslandi: Þernur og svartfuglar. *Náttúrufræðingurinn* 57: 137-143.

Pétursson, G. 1995. Sedrustoppa á Íslandi. *Bliki* 16: 7-10.

Pétursson, G. 1996. Tregadúfa í Vestmannaeyjum. *Bliki* 17: 27-28.

Pétursson, G. & Ólafsson, E. 1985. Sjaldgæfir fuglar á Íslandi 1983. *Bliki* 4: 13-39.

Pétursson, G. & Ólafsson, E. 1989. Sjaldgæfir fuglar á Íslandi 1987. *Bliki* 8: 15-46.

Pétursson, G. & Þráinsson, G. 1999. Sjaldgæfir fuglar á Íslandi fyrir 1981. *Fjölrit Náttúrufræðistofnunar* 37. Reykjavík.

Pétursson, G., Þráinsson, G. & Ólafsson, E. 1991. Sjaldgæfir fuglar á Íslandi 1988. *Bliki* 10: 15-50.

Picozzi, N. 1971. Evening Grosbeak on St Kilda: a species new to Britain and Ireland. *British Birds* 64: 189-194.

Pineau, O., Kayser, Y., Sall, M., Gueye, A. & Hafner, H. 2001. The Kelp Gull at Banc d'Arguin - a new Western Palearctic bird. *Birding World* 14: 110-111.

Pinguinhas, M. 2006. A pale morph Trinidade Petrel in the Azores. *Birding World* 19: 210-211.

Piskunov, V. V. & Antonchikov, A. N. 2007. Flock of 14 Relict Gulls in southern Russia and north-western Kazakhstan in May 2000. *Dutch Birding* 29: 94-95.

Pope, M., Foster, B. & Fagel, P. 2006. The Forest Wagtail in Kuwait – a new Western Palearctic bird. *Birding World* 19: 482-483.

Preston, K. 1981. Twenty-eighth Irish Bird Report, 1980. *Irish Birds* 2: 87-122.

Preston, K. 1989. Gray Catbird in Co. Cork: new to Britain and Ireland. *British Birds* 82: 1-3.

Proctor, B. & Donald, C. 2003. Yellow-headed Blackbirds in Britain and Europe. *Birding World* 16: 69-81.

Rabbitts, B. 1999. The Mourning Dove on the Outer Hebrides. *Birding World* 12: 453.

Rabbitts, B. 2007. The Mourning Dove on the Outer Hebrides. *Birding World* 20: 456-458.

Rabbitts, B. 2008. Mourning Dove on North Uist: new to Britain. *British Birds* 101: 26-30.

Ramadan-Jaradi, G., Bara, T. & Ramadan-Jaradi, M. 2008. Revised checklist of the birds of Lebanon 1999-2007. *Sandgrouse* 30: 22-69.

Rannila, V.-P. 2003. The Oriental Plover in Finland - a new Western Palearctic bird. *Birding World* 16: 209.

Rariteettikomitea 1984. Ennen Rariteettikomitean perustamista julkaistujen harvinaisuushavaintojen tarkistus. *Lintumies* 19: 76-82.

Rasmussen, P. A. F. 1998. Sjældne fugle i Danmark og Grønland i 1997. *Dansk Ornitologisk Forening Tidsskrift* 92: 253-267.

Ree, V. 1976. Rapport fra NNSK's virksomhet april 1975 – april 1976. *Sterna* 15: 179-197.

Ree, V. 2009. The Brown-headed Cowbird in Norway. *Birding World* 22: 260.

Reeber, S., Frémont, J.-Y., Flitti, A. & CHN 2008. Les oiseaux rares en France en 2006-2007. *Ornithos* 15: 313-355.

REFERENCES

Riddiford, N. 1983. Sandhill Crane: new to Britain. *British Birds* 76: 105-109.

Riddington, R. & Reid, J. 2000. Lesser Frigatebird in Israel: new to the Western Palearctic. *British Birds* 93: 22-27.

Robertson, I. S. 1975. Ovenbird in Shetland: a species new to Britain and Ireland. *British Birds* 68: 453-455.

Rogers, M. J. & Rarities Committee 1978. Report on rare birds in Great Britain in 1977. *British Birds* 71: 481-532.

Rogers, M. J. & Rarities Committee 1979. Report on rare birds in Great Britain in 1978. *British Birds* 72: 503-549.

Rogers, M. J. & Rarities Committee 1980. Report on rare birds in Great Britain in 1979. *British Birds* 73: 491-534.

Rogers, M. J. & Rarities Committee 1981. Report on rare birds in Great Britain in 1980. *British Birds* 74: 453-495.

Rogers, M. J. & Rarities Committee 1982. Report on rare birds in Great Britain in 1981. *British Birds* 75: 482-533.

Rogers, M. J. & Rarities Committee 1983. Report on rare birds in Great Britain in 1982. *British Birds* 76: 476-529.

Rogers, M. J. & Rarities Committee 1984. Report on rare birds in Great Britain in 1983. *British Birds* 77: 506-562.

Rogers, M. J. & Rarities Committee 1985. Report on rare birds in Great Britain in 1984. *British Birds* 78: 529-589.

Rogers, M. J. & Rarities Committee 1986. Report on rare birds in Great Britain in 1985. *British Birds* 79: 526-588.

Rogers, M. J. & Rarities Committee 1987. Report on rare birds in Great Britain in 1986. *British Birds* 80: 516-571.

Rogers, M. J. & Rarities Committee 1988. Report on rare birds in Great Britain in 1987. *British Birds* 81: 535-596.

Rogers, M. J. & Rarities Committee 1989. Report on rare birds in Great Britain in 1988. *British Birds* 82: 505-563.

Rogers, M. J. & Rarities Committee 1990. Report on rare birds in Great Britain in 1989. *British Birds* 83: 439-496.

Rogers, M. J. & Rarities Committee 1991. Report on rare birds in Great Britain in 1990. *British Birds* 84: 449-505.

Rogers, M. J. & Rarities Committee 1992. Report on rare birds in Great Britain in 1991. *British Birds* 85: 507-554.

Rogers, M. J. & Rarities Committee 1993. Report on rare birds in Great Britain in 1992. *British Birds* 86: 447-540.

Rogers, M. J. & Rarities Committee 1994. Report on rare birds in Great Britain in 1993. *British Birds* 87: 503-571.

Rogers, M. J. & Rarities Committee 1995. Report on rare birds in Great Britain in 1994. *British Birds* 88: 493-558.

Rogers, M. J. & Rarities Committee 1996. Report on rare birds in Great Britain in 1995. *British Birds* 89: 481-531.

Rogers, M. J. & Rarities Committee 1997. Report on rare birds in Great Britain in 1996. *British Birds* 90: 453-522.

Rogers, M. J. & Rarities Committee 1998. Report on rare birds in Great Britain in 1997. *British Birds* 91: 455-517.

Rogers, M. J. & Rarities Committee 1999. Report on rare birds in Great Britain in 1998. *British Birds* 92: 554-609.

Rogers, M. J. & Rarities Committee 2000. Report on rare birds in Great Britain in 1999. *British Birds* 93: 512-567.

Rogers, M. J. & Rarities Committee 2002. Report on rare birds in Great Britain in 2001. *British Birds* 95: 476-528.

Rogers, M. J. & Rarities Committee 2003. Report on rare birds in Great Britain in 2002. *British Birds* 96: 542-609.

Rogers, M. J. & Rarities Committee 2004. Report on rare birds in Great Britain in 2003. *British Birds* 97: 558-625.

Rogers, M. J. & Rarities Committee 2005. Report on rare birds in Great Britain in 2004. *British Birds* 98: 628-694.

Roselaar, C. S. 2006. The boundaries of the Palearctic region. *British Birds* 99: 602-618.

Round, P. D. 1996. Long-toed Stint in Cornwall: the first record for the Western Palearctic. *British Birds* 89: 12-24.

Rountree, P. 1977. Siberian Blue Robin: new to Europe. *British Birds* 70: 361-365.

Rowlands, A., Kidner, P. & Condon, P. 2010. From the Rarities Committee's files: Yellow-nosed Albatross: new to Britain. *British Birds* 103: 376-384.

Ruttledge, R. F. 1962. *9th Irish Bird Report 1961*. Dublin.

Ruttledge, R. F. 1971. *18th Irish Bird Report 1970*. Dublin.

Ryabitsev, V. K. 2008. *Ptitsy Urala, Priural'ya i Zapadnoy Sibiri: Spravochnik-opredlitel'*. Third edition. Ekaterinburg.

Rylands, K. 2008. Long-billed Murrelet in Devon: new to Britain. *British Birds* 101: 131-136.

Salvadori 1912-13. Singolare cattura di una specie orientale del genere *Ardetta*. Nuova per L'Italia e per L'Europa. *Rivista italiana di ornitologia* 2: 86-88.

Sanders, E., Lilipaly, S. J. & Ebels, E. B. 1998. Stekelstaartgierzwaluw op Walcheren in mei 1996. *Dutch Birding* 20: 168-172.

Sanders, J. 2007. The Glaucous-winged Gull in Gloucestershire: a new British bird. *Birding World* 20: 13-19.

Sanders, J. 2010. Glaucous-winged Gull in Gloucestershire: new to Britain. *British Birds* 103: 53-59.

Sangster, G., van den Berg, A. B., van Loon, A. J. & Roselaar, C. S. 2003. Dutch avifaunal list: taxonomic changes in 1999-2003. *Ardea* 91: 281-287.

Sangster, G., van den Berg, A. B., van Loon, A. J. & Roselaar, C. S. 2009. Dutch avifaunal list: taxonomic changes in 2004-2008. *Ardea* 97: 373-381.

Sangster, G., Hazevoet, C. J., van den Berg, A. B., Roselaar, C. S. & Sluys, R. 1999. Dutch avifaunal list: species concepts, taxonomic instability, and taxonomic changes in 1977-1998. *Ardea* 87: 139-165.

Sapir, N. & Israeli Rarities and Distribution Committee 2007. Bulletin 6:01 on *Rare Birds in Israel*: Updated National Bird Species List. Website http://www.israbirding.com/irdc/bulletins/bulletin_6/.

Sapsford, A. 1990. Mourning Dove - A new Western Palearctic bird. *Birding World* 3: 64.

Sapsford, A. 1996. Mourning Dove in the Isle of Man: new to the Western Palearctic. *British Birds* 89: 157-161.

Saunders, D. & Saunders, S. 1992. Blackburnian Warbler: new to the Western Palearctic. *British Birds* 85: 337-343.

Scharringa, C. J. G. & Winkelman, J. E. 1984. Zeldzame en schaarse vogels in Nederland in 1982. *Limosa* 57: 17-26.

van der Schot, W. E. M. 1989. Atlantic Petrel in Israel in April 1989. *Dutch Birding* 11: 170-172.

Scott, M. 2002. A Brown Skua on the Isles of Scilly – the first for Europe? *Birding World* 15: 383-386.

Scott, M. 2004. The Redhead on the Outer Hebrides - the first female for the Western Palearctic. *Birding World* 17: 59.

Shaw, D. 2003a. The Thick-billed Warbler on Fair Isle. *Birding World* 16: 206-208.

Shaw, D. 2003b. The Savannah Sparrow on Fair Isle. *Birding World* 16: 423-247.

Shaw, D. 2004a. The Chestnut-eared Bunting on Fair Isle - a new Western Palearctic bird. *Birding World* 17: 415-419.

Shaw, D. 2004b. The Rufous-tailed Robin on Fair Isle - a new Western Palearctic bird. *Birding World* 17: 420-423.

Shaw, D. N. 2006. Rufous-tailed Robin on Fair Isle: new to Britain. *British Birds* 99: 236-241.

Shaw, D. N. 2008. Chestnut-eared Bunting on Fair Isle: new to Britain. *British Birds* 101: 235-240.

Shaw, D. 2009. The Brown-headed Cowbird on Fair Isle – the second Scottish record. *Birding World* 22: 200-203.

Shirihai, H. 1996. *The birds of Israel*. London.

Shirihai, H. 1999. Fifty species new to Israel, 1979-1998: their discovery and documentation, with tips on identification. *Sandgrouse* 21: 45-105.

Shirihai, H. 2000. The first Southern Pochard *Netta erythrophthalma* in Israel and the Western Palearctic. *Sandgrouse* 22: 131-133.

Shirihai, H. & Alström, P. 1990. Identification of Hume's Short-toed Lark. *British Birds* 83: 262-272.

Shirihai, H. & Golan, Y. 1994. First records of Long-tailed Shrike *Lanius schach* in Israel and Turkey. *Sandgrouse* 16: 36-40.

Shirihai, H., Jonsson, A. & Sebba, N. 1987. Brown-headed Gull in Israel in May 1985. *Dutch Birding* 9: 120-122.

Shirihai, H. & Sinclair, I. 1994. An unidentified shearwater at Eilat. *Birding World* 7:274-278.

Shirihai, H., Sinclair, I. & Colston, P. R. 1995. A new species of *Puffinus* shearwater from the western Indian Ocean. *Bulletin of the British Ornithologists' Club* 115: 75-87.

Shochat, E., Sapir, N. & Israeli Rarities and Distribution Committee 2004. Bulletin 3:01 on *Rare Birds in Israel (1995 – 2001)*. Website http://www.israbirding.com/irdc/bulletins/bulletin_3/.

Siblet, J.-P. & Spanneut, L. 1998. The Willet in Vendée, France. *Birding World* 11: 386.

Sim, G. 1880. Esquimaux Curlew in Kincardineshire. *The Zoologist* 4 (third series): 515-516.

Simon, P. 1965. Synthèse de l'avifaune du massif montagneux du Tibesti, et distribution géographique de ces espèces en Afrique du nord et environs. *Gerfaut* 55: 26-71.

Slack, R. 2009. *Rare birds Where and When. An Analyses of Status & Distribution in Britain and Ireland*. Volume 1. York.

Smaldon, R. 1990. Wilson's Warbler: new to the Western Palearctic. *British Birds* 83: 404-408.

de Smet, G. 1996. De eerste Geelbrauwgors *Emberiza chrysophrys* in het Westpalearctisch gebied: de vergeten vogel van oktober 1966. *Oriolus* 62: 92-94.

de Smet, G., Pollet, J. & BAHC 1996. Zeldzame vogels in België in 1994. *Oriolus* 62: 37-54.

Smiddy, P. & O'Sullivan, O. 1996. Forty-third Irish Bird Report, 1995. *Irish Birds* 5: 445-474.

Smith, F. R. & Rarities Committee 1967. Report on rare birds in Great Britain in 1966. *British Birds* 60: 309-338.

Smith, F. R. & Rarities Committee 1968. Report on rare birds in Great Britain in 1967. *British Birds* 61: 329-365.

Smith, F. R. & Rarities Committee 1969. Report on rare birds in Great Britain in 1968. *British Birds* 62: 457-492.

Smith, F. R. & Rarities Committee 1970. Report on rare birds in Great Britain in 1969. *British Birds* 63: 267-293.

Smith, F. R. & Rarities Committee 1972. Report on rare birds in Great Britain in 1971. *British Birds* 65: 322-354.

Smith, F. R. & Rarities Committee 1974. Report on rare birds in Great Britain in 1973. *British Birds* 67: 310-348.

Smith, J. P. & Israeli Rarities and Distribution Committee 2001. Bulletin 1:02 on *Rare Birds in Israel (1995 – 2001)*. Website http://www.israbirding.com/irdc/bulletins/bulletin_1.2/.

Smith, J. P. & Israeli Rarities and Distribution Committee 2002. Bulletin 2:01 on *Rare Birds in Israel (1995 – 2002)*. Website http://www.israbirding.com/irdc/bulletins/bulletin_2/.

Smith, P. 1996. The Cedar Waxwing in Nottingham - a new British bird. *Birding World* 9: 70-73.

Soldaat, E., Leopold, M. F., Meesters, E. H. & Robertson, C. J. R. 2009. Albatross mandible at archeological site in Amsterdam, the Netherlands, and WP records of *Diomedea* albatrosses. *Dutch Birding* 31: 1-16.

Sondbø, S. 1992. The Willet in Norway. *Birding World* 5: 458-460.

O' Sullivan, J. & the Rarities Committee 1977. Report on rare birds in Great Britain in 1976. *British Birds* 70: 405-453.

Steffen, B. 2010. African Openbill at Luxor, Egypt, in May 2009. *Dutch Birding* 32: 254-256.

Stenning, J. & Hirst, P. 1994. The Grey-tailed Tattler in Grampian - the second Western Palearctic record. *Birding World* 7: 469-472.

Stevenson, A. 2000. The Long-tailed Shrike on the Outer Hebrides - a new British bird. *Birding World* 13: 454-457.

Stevenson, A. 2005. Long-tailed Shrike: new to Britain. *British Birds* 98: 26-31.

Stjernberg, T. 1999. Der erste Fund der Elsterdohle in Europa: Anfang Mai 1883 in Finnland. *Blätter aus dem Naumann-Museum* 18: 54-60.

Stühmer, F 2005. Flycatcher on Helgoland, Germany, in August 1982 re-identified as Dark-sided Flycatcher. Dutch Birding 27: 204-205.

O'Sullivan, D. 1996. The Long-toed Stint in County Cork - the first for Ireland. *Birding World* 9: 224-225.

O'Sullivan, O. & Smiddy, P. 1987. Thirty-fourth Irish Bird Report, 1986. *Irish Birds* 3: 455-490.

O'Sullivan, O. & Smiddy, P. 1989. Thirty-sixth Irish Bird Report, 1988. *Irish Birds* 4: 79-114.

O'Sullivan, O. & Smiddy, P. 1991. Thirty-eighth Irish Bird Report, 1990. *Irish Birds* 4: 423-462.

Sultana, J. 1968. The occurrence of the American Kestrel (*Falco sparverius*) in Malta. *Maltese Ornithological Society Quarterly Bulletin* 2:9.

Sultana, J. & Gauci, C. 1984-85. Two new species for Malta - Red-eyed Vireo and Chestnut Bunting. *Il-Merill* 23: 11.

Symens, P. & Spanoghe, G. 2007. Nieuw voor België: Katvogel *Dumetella carolinensis* te Kallo in december 2006. *Natuur.oriolus* 73: 13-16.

Tamás, Z. 2008. Az Amuri Vércse *(Falco amurensis)* elsö elöfordulása Magyarországon. *Aquila* 114-115: 71-73.

Taylor, A. M. 1981. American Kestrel: new to Britain and Ireland. *British Birds* 74: 199-203.

Taylor, B. & van Perlo, B. 1998. Rails. *A guide to the Rails, Crakes, Gallinules and Coots of the World*. Mountfield.

Taylor, G., Garner, M. & McLoughlin, J. 2007. The Pacific Diver in North Yorkshire – a new Western Palearctic bird. *Birding World* 20: 20-25.

Thévenot, M., Vernon, R. & Bergier, P. 2003. *The birds of Morocco*. Tring.

Thiede, W. 1966. Ein vergessener Katzenvogel. *Ornithologische Mitteilungen* 18: 205.

Thorpe, R. I 1995. Grey-tailed Tattler in Wales: new to Britain and Ireland. *British Birds* 88: 255-262.

Þórisson, S. G. 2007. Barrþröstur sést í annað sinn í Evrópu. *Bliki* 28: 59-60.

Þráinsson, G. 1997. Palm Warbler and Cerulean Warbler in Iceland - new to the Western Palearctic. *Birding World* 10: 392-393.

Þráinsson, G. 1998. Daggarskríkjur finnast á Íslandi. *Bliki* 19: 43-44.

Þráinsson, G. & Pétursson, G. 1997. Sjaldgæfir fuglar á Íslandi 1995. *Bliki* 18: 23-30.

Þráinsson, G., Pétursson, G. & Kolbeinsson, Y. 2009. Sjaldgæfir fuglar á Íslandi 2006. *Bliki* 30: 27-46.

Þráinsson, G., Pétursson, G. & Ólafsson, E. 1995. Sjaldgæfir fuglar á Íslandi 1993. *Bliki* 15: 21-51.

Trnka, A. & Matousek, B. 1996. Occurrence of Atlantic Petrel in Western Palearctic. *Dutch Birding* 18: 309-310.

Tye, A. 1994. A description of the Middle Eastern black morph of Mourning Wheatear *Oenanthe lugens* from museum specimens. *Sandgrouse* 16: 28-31.

Undeland, P. & Leuzinger, H. 1992. Seltene Vogelarten und ungewöhnliche Vogelbeobachtungen in der Schweiz im Jahre 1991. *Der Ornithologische Beobachter* 89: 253-265.

Unger, U. 1979. Sällsynta fåglar i Sverige 1977 – rapport från SOF:s raritetskommitté. *Vår Fågelvärld* 38: 39-46.

Ussher, R. J. & Warren, R 1900. *The Birds of Ireland*. London.

Valverde, J. A. 1957. *Aves del Sahara Español*. Madrid.

Vandegehuchte, M. & BAHC 2008. Zeldzame vogels in België in 2006. *Natuur.oriolus* 74: 45-51.

Vasamies, H. 1997. Little Whimbel at Åland. *Alula* 3: 38-40.

Velaco, D. L. 2010. Identification of the first Pacific Diver for Spain. *Birding World* 23: 14-19.

Velmala, W., Clarke, T., Lindroos, R. & Nikkinen, L. 2002. A Dwarf Bittern on the Canary Islands - the fourth Western Palearctic record. *Birding World* 15: 392-393.

Vinicombe, K. & Cottridge, D. M. 1996. *Rare Birds in Britain and Ireland. A photographic Record*. London.

van Vliet, P. J. & Ebels, E. B. 2007. Geelkoptroepiaal op Texel in mei-juni 1982. *Dutch Birding* 29: 232.

van der Vliet, R. E., van der Laan, J., Berlijn, M. & CDNA 2006. Rare birds in the Netherlands in 2005. *Dutch Birding* 28: 345-365.

van der Vliet, R E, van der Laan, J., Berlijn, M & CDNA 2007. Rare birds in the Netherlands in 2006. Dutch Birding 29: 347-374.

van der Vliet, R. E., van der Laan, J. & CDNA 2001. Rare birds in the Netherlands in 2000. *Dutch Birding* 23: 315-347.

van der Vliet, R. E., van der Laan, J. & CDNA 2002. Rare birds in the Netherlands in 2001. *Dutch Birding* 24: 325-349.

van der Vliet, R. E., van der Laan, J. & CDNA 2003. Rare birds in the Netherlands in 2002. *Dutch Birding* 25: 361-384.

van der Vliet, R. E., van der Laan, J. & CDNA 2004. Rare birds in the Netherlands in 2003. *Dutch Birding* 26: 359-384.

van der Vliet, R. E., van der Laan, J. & CDNA 2005. Rare birds in the Netherlands in 2004. *Dutch Birding* 27: 367-394.

Vonk, H. & van IJzendoorn, E. J. 1988. Geelbrauwgors op Schiermonnikoog in oktober 1982. *Dutch Birding* 10: 127-130, 1988.

Votier, S. C., Bearhop, S., Newell, R. G., Orr, K., Furness, R. W. & Kennedy, M. 2004. The first record of Brown Skua *Catharacta antarctica* in Europe. *Ibis* 146: 95-102.

Votier, S. C., Kennedy, M., Bearhop, S., Newell, R. G., Griffiths, K., Whitaker, H., Ritz, M. S. & Furness, R. W. 2007. Supplementary DNA evidence fails to confirm presence of Brown Skuas *Stercorarius antarctica* in Europe: a retraction of Votier et al (2004). *Ibis* 149: 619-621.

Wagstaff, W. 1990. Tree Swallow on Scilly - A new Western Palearctic bird. *Birding World* 3: 199-201.

Walbridge, G., Small, B. & McGowan, R. 2003. Ascension Frigatebird on Tiree - new to the Western Palearctic. *British Birds* 96: 58-73.

Waldon, J. 1994. Ancient Murrelet in Devon: new to the Western Palearctic. *British Birds* 87: 307-310.

Walker, R. J. & Gregory, J. 1987. Little Whimbrel in Norfolk. *British Birds* 80: 494-497.

Waller, C. S. 1970. Rufous-sided Towhee on Lundy. *British Birds* 63: 147-149.

Ward, N. 1987. Ovenbird in Devon. *British Birds* 80: 500-502.

Wassink, A. & Oreel, G. J. 2007. *The birds of Kazakhstan*. De Cocksdorp.

Watmough, N. 1988. Yellow-bellied Sapsucker in County Cork. *Birding World* 1: 392-394.

van Welie, L. 2000. Lesser Frigatebird at Eilat, Israel, in May 1999. *Dutch Birding* 22: 16-17.

Wiegant, W. M., de Bruin, A. & CDNA 1998. Rare birds in the Netherlands in 1996. *Dutch Birding* 20: 145-167.

Wiegant, W. M., de Bruin, A. & CDNA 1999. Rare birds in the Netherlands in 1997. *Dutch Birding* 21: 65-81.

Wiegant, W. M., Steinhaus, G. H. & CDNA 1994a. Rare birds in the Netherlands in 1992. *Dutch Birding* 16: 133-147.

Wiegant, W. M., Steinhaus, G. H. & CDNA 1994b. Zeldzame en schaarse vogels in Nederland in 1992. *Limosa* 67: 163-172.

Wiegant, W. M., Steinhaus, G. H. & CDNA 1995. Rare birds in the Netherlands in 1993. *Dutch Birding* 17: 89-101.

Wiegant, W. M., Steinhaus, G. H. & CDNA 1996a. Rare birds in the Netherlands in 1994. *Dutch Birding* 18: 105-121.

Wiegant, W. M., Steinhaus, G. H. & CDNA 1996b. Zeldzame en schaarse vogels in Nederland in 1993. *Limosa* 69: 13-22.

Wiegant, W. M., Steinhaus, G. H. & CDNA 1997. Rare birds in the Netherlands in 1995. *Dutch Birding* 19: 97-115.

Wilde, J. 1962. Fox Sparrow in Co. Down: a bird new to Britain and Ireland. *British Birds* 55: 560-562.

Williamson, K., Thom, V. M., Ferguson-Lees, I. J. & Axell, H. E. 1956. Thick-billed Warbler at Fair Isle: a new British bird. *British Birds* 49: 89-93.

Willmott, M. J. 1988. Blackburnian Warbler on Fair Isle - a new Western Palearctic bird. *Birding World* 1: 355-356.

Wilson, H. J. 1980. Ovenbird *Seirurus aurocapillus*: a new species to Ireland. *Irish Naturalist's Journal* 20: 125.

Wilson, K. 2008. The Alder Flycatcher in Cornwall – a new British bird. *Birding World* 21: 425-431.

Wing, S. 2000. The Blue-winged Warbler in County Cork - a new Western Palearctic bird. *Birding World* 13: 408-411.

Winkel, E. 2009. 'Southern skua' off La Palma, Canary Islands, in October 2005. *Dutch Birding* 31: 20-23.

Witherby, H. F 1907. Canadian Crane in Ireland. *British Birds* 1: 90-91.

Wright, G. 1987. Hudsonian Godwit in Devon. *British Birds* 80: 492-494.

Wright, J. 1994. The Yellow-browed Bunting on Scilly. *Birding World* 7: 410-411.

Wright, M. 2009. The Tufted Puffin in Kent – a new British bird. *Birding World* 22: 374-375.

Yates, B. 2010. Least Tern in East Sussex: new to Britain and the Western Palearctic. *British Birds* 103: 339-347.

Yates, B. & Taffs, H. 1990. Least Tern in East Sussex - A new Western Palearctic bird. *Birding World* 3: 197-199.

Zuyderduyn, C. 2008. Kroonboszanger in Katwijk aan Zee in oktober 2007. *Dutch Birding* 30: 155-158.

APPENDIX

NEW SPECIES FOR THE WP SINCE 2009

The following three species have been recorded since 2009 (that is, after the end of the period covered by this book, 1800–2008). All three records are documented by photographs.

African Openbill *Anastomus lamelligerus*

1 record

Egypt (1)
1 • 26 May 2009, Crocodile Island, Luxor [Gantlett 2010; Steffen 2010]

Asian Koel *Eudynamys scolopaceus*

1 record

Kuwait (1)
1 • 22–28 February 2009, Abdali farms [Gantlett 2010]

Ashy Drongo *Dicrurus leucophaeus*

1 record

Kuwait (1)
1 • 03–04 April 2010, Jahra farms [van den Berg & Haas 2010c]

INDEX

A

Acadian Flycatcher   123
Acrocephalus orientalis   148
Aenigmatolimnas marginalis   70
Aethia cristatella   108
Aethia psittacula   109
African Crake   72
African Fish Eagle   64
African Openbill   237
African Palm Swift   117
Agropsar sturninus   157
Alder Flycatcher   124
Aleutian Tern   102
American Kestrel   66
American Woodcock   84
Amur Falcon   68
Amur Wagtail   173
Anas capensis   32
Anas erythrorhyncha   32
Anastomus lamelligerus   237
Ancient Murrelet   107
Anous stolidus   104
Anser rossii   19
Ant Chat   171
Apus pacificus   116
Ardea melanocephala   61
Ardeola grayii   58
Ascension Frigatebird   52
Ashy Drongo   238
Asian Brown Flycatcher   165
Asian Koel   238
Atlantic Petrel   44
Atlantic Yellow-nosed Albatross   35
Aythya americana   25

B

Bald Eagle   64
Banded Martin   137
Bay-breasted Warbler   203
Bermuda Petrel   42
Blackburnian Warbler   198
Black Heron   61
Black-capped Petrel   43
Black-headed Heron   61
Black-headed Lapwing   80
Black-throated Blue Warbler   195
Black-throated Green Warbler   197
Blue-winged Warbler   191
Bombycilla cedrorum   149
Brachyrhampus perdix   105
Broad-billed Roller   118
Brown Noddy   104
Brown Thrasher   154
Brown-headed Cowbird   188
Brown-headed Gull   96

C

Calandrella acutirostris   136
Calidris subminuta   81
Calonectris leucomelas   45
Canada Warbler   208
Cape Gull   99
Cape May Warbler   199
Cape Petrel   40
Cape Teal   32
Catharus fuscescens   162
Cedar Waxwing   149
Cerulean Warbler   194
Charadrius tricollaris   78
Charadrius veredus   79
Chestnut Bunting   185
Chestnut-eared Bunting   182
Chestnut-headed Sparrow-Lark   136
Chestnut-sided Warbler   193
Chondestes grammacus   178
Chrysococcyx caprius   114
Cinnyris asiaticus   159
Clamator jacobinus   114
Colaptes auratus   119
Columba eversmanni   110
Corvus albus   134
Corvus dauuricus   132

INDEX

Cotton Pygmy Goose   24
Crested Auklet   108
Crex egregia   72
Croicocephalus brunnicephalus   96
Cypsiurus parvus   117

D

Daption capense   40
Daurian Jackdaw   132
Daurian Redstart   170
Daurian Starling   157
Dendrocygna javanica   18
Dendroica caerulescens   195
Dendroica castanea   203
Dendroica cerulea   194
Dendroica fusca   198
Dendroica magnolia   200
Dendroica palmarum   202
Dendroica pensylvanica   193
Dendroica tigrina   199
Dendroica virens   197
Dendronanthus indicus   172
Dickcissel   177
Dicrurus leucophaeus   238
Dideric Cuckoo   114
Diomedea dabbenena   38
Dumetella carolinensis   155
Dwarf Bittern   57

E

Eastern Crowned Warbler   142
Eastern Phoebe   122
Eastern Towhee   177
Egretta ardesiaca   61
Egretta caerulea   59
Egretta tricolor   60
Elanoides forficatus   63
Emberiza chrysophrys   183
Emberiza fucata   182
Emberiza rutila   185
Empidonax alnorum   124

Empidonax minimus   125
Empidonax virescens   123
Eremopterix signatus   136
Eskimo Curlew   90
Ethiopian Swallow   141
Eudynamys scolopaceus   238
Eurystomus glaucurus   118
Evening Grosbeak   174

F

Falco amurensis   68
Falco sparverius   66
Ficedula mugimaki   171
Flesh-footed Shearwater   46
Forest Wagtail   172
Fork-tailed Flycatcher   125
Fork-tailed Swift   116
Fox Sparrow   181
Fratercula cirrhata   109
Fregata aquila   52
Fregata ariel   53

G

Gallinago megala   83
Gallinula angulata   74
Gavia pacifica   33
Glaucous-winged Gull   97
Golden-winged Warbler   190
Gray's Grasshopper Warbler   145
Grey Catbird   155
Grey-tailed Tattler   91
Grus canadensis   75

H

Haliaeetus leucocephalus   64
Haliaeetus vocifer   64
Hesperiphona vespertina   174
Hirundo aethiopica   141
Hooded Vulture   65
Hooded Warbler   207
Hudsonian Godwit   85

241

Hume's Short-toed Lark   136
Hylocichla mustelina   161

I

Iduna aedon   146
Indian Pond Heron   58
Ixobrychus eurhythmus   56
Ixobrychus exilis   54
Ixobrychus sturmii   57
Ixoreus naevius   160

J

Jacobin Cuckoo   114

L

Lanius schach   129
Lark Sparrow   178
Larus dominicanus   99
Larus glaucescens   97
Larus relictus   96
Larus schistisagus   101
Least Bittern   54
Least Flycatcher   125
Least Tern   103
Leptoptilos crumeniferus   62
Lesser Frigatebird   53
Lesser Moorhen   74
Lesser Whistling Duck   18
Limosa haemastica   85
Little Blue Heron   59
Little Curlew   88
Locustella fasciolata   145
Long-billed Murrelet   105
Long-tailed Shrike   129
Long-toed Stint   81
Louisiana Waterthrush   206
Luscinia cyane   168
Luscinia sibilans   167

M

Macronectes giganteus   39

Magnolia Warbler   200
Marabou Stork   62
Masked Booby   50
Melanitta deglandi   29
Mimus polyglottos   152
Molothrus ater   188
Motacilla leucopsis   173
Mourning Dove   111
Mugimaki Flycatcher   171
Muscicapa dauurica   165
Myrmecocichla aethiops   171

N

Necrosyrtes monachus   65
Netta erythrophthalma   24
Nettapus coromandelianus   24
Northern Flicker   119
Northern Mockingbird   152
Numenius borealis   90
Numenius minutus   88

O

Oenanthe picata   171
Onychoprion aleuticus   102
Oriental Plover   79
Oriental Reed Warbler   148
Ovenbird   203

P

Pacific Loon   33
Palm Warbler   202
Parakeet Auklet   109
Passerculus sandwichensis   179
Passerella iliaca   181
Phaethon lepturus   48
Philadelphia Vireo   128
Phoenicurus auroreus   170
Phylloscopus coronatus   142
Phylloscopus neglectus   144
Pied Crow   134
Pipilo erythrophthalmus   177

INDEX

Piranga rubra   176
Plain Leaf Warbler   144
Plectropterus gambensis   23
Progne subis   140
Pterodroma arminjoniana   41
Pterodroma cahow   42
Pterodroma hasitata   43
Pterodroma incerta   44
Pterodroma mollis   42
Puffinus bailloni   47
Puffinus carneipes   46
Puffinus pacificus   46
Purple Martin   140
Purple Sunbird   159

R

Red-billed Teal   32
Red-breasted Nuthatch   151
Red-footed Booby   49
Redhead   25
Regulus calendula   135
Relict Gull   96
Riparia cincta   137
Ross's Goose   19
Ruby-crowned Kinglet   135
Rufous-tailed Robin   167

S

Sandhill Crane   75
Savannah Sparrow   179
Sayornis phoebe   122
Scolopax minor   84
Seiurus aurocapilla   203
Seiurus motacilla   206
Shy Albatross   37
Siberian Blue Robin   168
Sitta canadensis   151
Slaty-backed Gull   101
Soft-plumaged Petrel   42
Somateria fischeri   28
South Polar Skua   94

Southern Giant Petrel   39
Southern Pochard   24
Spectacled Eider   28
Sphyrapicus varius   120
Spiza americana   177
Spur-winged Goose   23
Stercorarius maccormicki   94
Sternula antillarum   103
Streaked Shearwater   45
Striped Crake   70
Sula dactylatra   50
Sula sula   49
Summer Tanager   176
Swallow-tailed Kite   63
Swinhoe's Snipe   83
Synthliboramphus antiquus   107

T

Tachycineta bicolor   138
Tennessee Warbler   192
Thalassarche cauta   37
Thalassarche chlororhynchos   35
Thick-billed Warbler   146
Three-banded Plover   78
Tickell's Thrush   164
Toxostoma rufum   154
Tree Swallow   138
Tricolored Heron   60
Trindade Petrel   41
Tringa brevipes   91
Tringa semipalmata   92
Tristan Albatross   38
Tropical Shearwater   47
Tufted Puffin   109
Turdus unicolor   164
Tyrannus savana   125

V

Vanellus tectus   80
Variable Wheatear   171
Varied Thrush   160

Veery   162
Vermivora chrysoptera   190
Vermivora peregrina   192
Vermivora pinus   191
Vireo flavifrons   127
Vireo griseus   126
Vireo philadelphicus   128
Von Schrenck's Bittern   56

W

Wedge-tailed Shearwater   46
White-eyed Vireo   126
White-tailed Tropicbird   48
White-winged Scoter   29
Willet   92
Wilsonia canadensis   208
Wilsonia citrina   207
Wilsonia pusilla   208
Wilson's Warbler   208
Wood Thrush   161

X

Xanthocephalus xanthocephalus   189

Y

Yellow-bellied Sapsucker   120
Yellow-browed Bunting   183
Yellow-eyed Pigeon   110
Yellow-headed Blackbird   189
Yellow-throated Vireo   127

Z

Zenaida macroura   111